Statistical Group Comparison

WILEY SERIES IN PROBABILITY AND STATISTICS

Established by WALTER A. SHEWHART and SAMUEL S. WILKS

Editors: *David J. Balding, Peter Bloomfield, Noel A. C. Cressie,
Nicholas I. Fisher, Iain M. Johnstone, J. B. Kadane, Louise M. Ryan,
David W. Scott, Adrian F. M. Smith, Jozef L. Teugels;*
Editors Emeriti: *Vic Barnett, J. Stuart Hunter, David G. Kendall*

A complete list of the titles in this series appears at the end of this volume.

Statistical Group Comparison

TIM FUTING LIAO

A JOHN WILEY & SONS, INC., PUBLICATION

This book is printed on acid-free paper. ∞

Copyright ©2002 John Wiley & Sons, Inc. All rights reserved.

Published simultaneously in Canada.

No part of this publication may be reproduced, stored in a retrieval system or transmitted in any form or by any means, electronic, mechanical, photocopying, recording, scanning or otherwise, except as permitted under Section 107 or 108 of the 1976 United States Copyright Act, without either the prior written permission of the Publisher, or authorization through payment of the appropriate per-copy fee to the Copyright Clearance Center, 222 Rosewood Drive, Danvers, MA 01923, (978) 750-8400, fax (978) 750-4744. Requests to the Publisher for permission should be addressed to the Permissions Department, John Wiley & Sons, Inc., 605 Third Avenue, New York, NY 10158-0012, (212) 850-6011, fax (212) 850-6008, E-Mail: PERMREQ@WILEY.COM.

For ordering and customer service, call 1-800-CALL-WILEY

Library of Congress Cataloging-in-Publication Data Is Available

ISBN 0-471-38646-4

Printed in the United States of America

10 9 8 7 6 5 4 3 2 1

To Athena and Keiko
May they always compare to discover

Contents

Preface xiii

1. Introduction 1

    1.1 Rationale for Statistical Comparison, 1
    1.2 Comparative Research in the Social Sciences, 3
    1.3 Focus of the Book, 5
    1.4 Outline of the Book, 5
        1.4.1 Chapter 2—Statistical Foundation for Comparison, 6
        1.4.2 Chapter 3—Comparison in Linear Regression, 6
        1.4.3 Chapter 4—Nonparametric Comparison, 6
        1.4.4 Chapter 5—Comparing Rates, 6
        1.4.5 Chapter 6—Comparison in Generalized Linear Models, 6
        1.4.6 Chapter 7—Additional Topics of Comparison in Generalized Linear Models, 7
        1.4.7 Chapter 8—Comparison in Structural Equation Modeling, 7
        1.4.8 Chapter 9—Comparison with Categorical Latent Variables, 8
        1.4.9 Chapter 10—Comparison in Multilevel Analysis, 8
        1.4.10 Summary, 8

2. Statistical Foundation for Comparison 9

    2.1 A System for Statistical Comparison, 9
    2.2 Test Statistics, 10
        2.2.1 The χ^2 Test, 10
        2.2.2 The t-Test, 12
        2.2.3 The F-test, 13

2.2.4 The Likelihood Ratio Test, 14
2.2.5 The Wald Test, 15
2.2.6 The Lagrange Multiplier Test, 17
2.2.7 A Summary Comparison of LRT, WT, and LMT, 18
2.3 What to Compare? 19
2.3.1 Comparing Distributions, 19
2.3.2 Comparing Data Structures, 19
2.3.3 Comparing Model Structures, 19
2.3.4 Comparing Model Parameters, 20

3. Comparison in Linear Models   21

3.1 Introduction, 21
3.2 An Example, 22
3.3 Some Preliminary Considerations, 23
3.4 The Linear Model, 23
3.5 Comparing Two Means, 24
3.6 ANOVA, 25
3.7 Multiple Comparison Methods, 27
  3.7.1 Least Significance Difference Test, 27
  3.7.2 Tukey's Model, 27
  3.7.3 Scheffé's Method, 28
  3.7.4 Bonferroni's Method, 29
3.8 ANCOVA, 30
3.9 Multiple Linear Regression, 30
3.10 Regression Decomposition, 33
  3.10.1 Rationale, 33
  3.10.2 Algebraic Presentation, 34
  3.10.3 Interpretation, 34
  3.10.4 Extension to Multiple Regression, 35
3.11 Which Linear Method to Use? 36

4. Nonparametric Comparison   37

4.1 Nonparametic Tests, 37
  4.1.1 Kolmogorov–Smirnov Two-Sample Test, 38
  4.1.2 Mann–Whitney U-Test, 38
4.2 Resampling Methods, 39
  4.2.1 Permutation Methods, 39
  4.2.2 Bootstrapping Methods, 42
4.3 Relative Distribution Methods, 44

5. Comparison of Rates   51

5.1 The Data, 51
5.2 Standardization, 52
    5.2.1 Direct Standardization, 52
    5.2.2 Indirect Standardization, 54
    5.2.3 Model-Based Standardization, 55
5.3 Decomposition, 58
    5.3.1 Arithmetic Decomposition, 58
    5.3.2 Model-Based Decomposition, 59

6. Comparison in Generalized Linear Models   62

6.1 Introduction, 62
    6.1.1 The Exponential Family of Distributions, 63
    6.1.2 The Link Function, 64
    6.1.3 Maximum Likelihood Estimation, 66
6.2 Comparing Generalized Linear Models, 68
    6.2.1 The Null Hypothesis, 68
    6.2.2 Comparisons Using Likelihood Ratio Tests, 68
    6.2.3 The Chow Test as a Special Case, 70
6.3 A Logit Model Example, 71
    6.3.1 The Data, 71
    6.3.2 The Model Comparison, 72
6.4 A Hazard Rate Model Example, 74
    6.4.1 The Model, 74
    6.4.2 The Data, 77
    6.4.3 The Model Comparison, 78
6.A Data Used in Section 6.4, 79

7. Additional Topics of Comparison in Generalized Linear Models   81

7.1 Introduction, 81
7.2 GLM for Matched Case–Control Studies, 81
    7.2.1 The $1:1$ Matched Study, 83
    7.2.2 The $1:m$ Design, 84
    7.2.3 The $n:m$ Design, 85
7.3 Dispersion Heterogeneity, 87
    7.3.1 The Data, 92
    7.3.2 Group Comparison with Heterogeneous Dispersion, 93
7.4 Bayesian Generalized Linear Models, 96
    7.4.1 Bayesian Inference, 96

7.4.2 An Example, 104
7.A The Data for the $n:m$ Design, 108

8. Comparison in Structural Equation Modeling — 111

8.1 Introduction, 111
8.2 Statistical Background, 112
  8.2.1 Notation and Specification, 112
  8.2.2 Identification, 115
  8.2.3 Estimation, 116
  8.2.4 Modification, 117
  8.2.5 Interpretation, 118
8.3 Mean and Covariance Structures, 119
8.4 Group Comparison in SEM, 121
  8.4.1 Equality of Multivariate Distributions, 122
  8.4.2 Equality of Covariance Matrices, 123
  8.4.3 Equality of Model Forms, 124
  8.4.4 Equality of Model Parameters, 125
8.5 An Example, 127
  8.5.1 Comparing Correlation Matrices, 127
  8.5.2 Comparing Covariance Structures and Multivariate Distributions, 130
  8.5.3 Comparing Mean and Covariance Structures, 132
8.A Examples of Computer Program Listings, 134
  8.A.1 An EQS Program for Comparing Correlation Matrices, 134
  8.A.2 A LISREL Program for Comparing Correlation Matrices, 136
  8.A.3 An EQS Program for Comparing Covariance Structures, 136
  8.A.4 A LISREL Program for Comparing Covariance Structures, 138
  8.A.5 An EQS Program for Comparing Mean and Covariance Structures, 139
  8.A.6 A LISREL Program for Comparing Mean and Covariance Structures, 140

9 Comparison with Categorical Latent Variables — 142

9.1 Introduction, 142
9.2 Latent Class Models, 143
9.3 Latent Trait Models, 147
9.4 Latent Variable Models for Continuous Indicators, 148

9.5 Causal Models with Categorical Latent Variables, 149
9.6 Comparison with Categorical Latent Variables, 151
- 9.6.1 Comparing Sampling Distributions, 152
- 9.6.2 Comparing Types and Patterns of Association between Variables, 152
- 9.6.3 Comparing Conditional Structure and Response Probabilities, 153
- 9.6.4 Comparing Latent Distributions and Conditional Probabilities, 153

9.7 Examples, 154
- 9.7.1 Comparison in Latent Class Analysis, 154
- 9.7.2 Comparison in a Path Model with Categorical Latent Variables, 158

9.A Software for Categorical Latent Variables, 161
- 9.A.1 MLLSA, 161
- 9.A.2 LAT, 161
- 9.A.3 PANMARK, 161
- 9.A.4 LCAG, 162
- 9.A.5 LEM, 162
- 9.A.6 Latent GOLD, 162
- 9.A.7 Mplus, 163
- 9.A.8 LATCLASS, TWOMISS, and POLYMISS, 163
- 9.A.9 lca.S and lcreg.sas, 163

9.B Computer Program Listings for the Examples, 163
- 9.B.1 A LEM Program for Comparing Sampling Zeros, 163
- 9.B.2 A LEM Program for Comparing LCMs Assuming Two-Way Interaction, 164
- 9.B.3 A LEM Program for Comparing LCMs Assuming Linear-by-Linear Association, 164
- 9.B.4 A LEM Program for Comparing LCMs Assuming Column-Effect RC-II Association, 165
- 9.B.5 A LEM Program for Comparing LCMs Assuming Complete Homogeneity, 165
- 9.B.6 A LEM Program for Comparing LCMs Assuming Complete Heterogeneity, 166
- 9.B.7 A LEM Program for Comparing LCMs Assuming Partial Heterogeneity, 166
- 9.B.8 A LEM Program for Comparing LCMs Assuming Partial Homogeneity, 167
- 9.B.9 Data for Example 2, 167
- 9.B.10 A LEM Program for Example 2 with Heterogeneous Groups, 169

9.B.11 A LEM Program for Example 2 with Homogeneous Measurement Parameters, 169
9.B.12 A LEM Program for Example 2 with Homogeneous Structural Parameters, 170
9.B.13 A LEM Program for Example 2 with Homogeneous Model Parameters, 171

10. Comparison in Multilevel Analysis   172

10.1 Introduction, 172
10.2 An Introduction to Multilevel Analysis, 173
    10.2.1 The Multilevel Setting, 173
    10.2.2 An Introductory Bibliography, 174
    10.2.3 Fixed versus Random Effects, 175
10.3 The Basics of the Linear Multilevel Model, 176
    10.3.1 The Basic Data Structure, 176
    10.3.2 Random Intercept Models, 177
    10.3.3 Random Coefficients Models, 179
    10.3.4 An Example, 180
    10.3.5 ANOVA with Random Effects, 181
    10.3.6 ANCOVA with Random Effects, 181
    10.3.7 Random Coefficient Models without Cross-Level Effects, 183
    10.3.8 Random Coefficient Models with Cross-Level Effects, 183
    10.3.9 Assumptions of the Linear Multilevel Model, 186
10.4 The Basics of the Generalized Linear Multilevel Model, 186
    10.4.1 A Random Coefficient Logit Model with Cross-Level Effects, 187
    10.4.2 A Random Coefficient Probit Model with Cross-Level Effects, 190
10.5 Group as an External Variable in Multilevel Analysis, 190
10.6 The Relation between Multilevel Analysis and Group Comparison, 191
    10.6.1 Bridging Fixed and Random Effects Models, 191
    10.6.2 An Example, 192
10.7 Multiple Membership Models, 193
10.8 Summary, 194
10.A Software for Multilevel Analysis, 195
    10.A.1 Special-Purpose Software, 195

>   10.A.2 General-Purpose Software, 195
>   10.A.3 Software for Other Special Purposes, 196
> 10.B SAS Program Listings for GLMM Examples, 197
>   10.B.1 Syntax for Producing the Logit Results in Table 10.11, 197
>   10.B.2 Syntax for Producing the Probit Results in Table 10.11, 197
>   10.B.3 Syntax for the Random Effects GLMM in Table 10.12, 198

References   **199**

Index   **207**

Preface

Doing science, be it natural or social, inevitably calls for comparison. Statistical methods are at the heart of such comparison, for they not only help us gain understanding of the world around us, but often define how our research is to be carried out.

The necessity to compare between groups is best exemplified by experiments, which have clearly defined statistical methods. However, more often than not in the social, political, education, and biomedical disciplines, true experiments are not possible. What complicates the matter more is a great deal of diversity in factors that are not independent of the outcome. At the same time these diverse groups form interesting basis for comparison. The awareness of and the interest in diversity, especially in the social, economical, and political sciences, have increased over the recent years, and this interest will only grow in the twenty-first century. The interest in as well as the need for comparative methods also is reflected in the many statistical methods developed for capturing the effect of this huge amount of diversity in composition, structure, and so on, when making comparison.

The aim of this book is to bring together a wide range of statistical methods for comparison developed over recent years and present them in a systematic manner. Thus the book covers a spectrum of topics from the simplest comparison of the means to more recently developed statistics, including Bayesian as well as hierarchical methods. Naturally, data analysts most often want to test a hypothesis of different outcomes, and the many tests described in the book reflect that desire. However, researchers are also interested in describing and explaining differences; for this reason the book covers some decomposition methods too, notably those for linear and loglinear rate comparisons.

Many of the examples in the book are from the social, political, and economic sciences. A few are from the biomedical sciences. But because of the range and the generality of the statistical methods covered, researchers across many disciplines—not just the social, political and biomedical sciences whose examples are use, will find the book of use. Both graduate students and researchers will find the book a convenient reference for many research situations when the need for comparison arises. The book can also be employed as a main text in a graduate seminar on statistical comparison, assuming its participants have familiarity with some form of linear, generalized linear, and latent variable models. In courses dealing with those specialty topics, the book may give supplementary coverage of comparative methods.

Although the book covers an extensive range of methods, it by no means includes all methods of comparison. If there is a bias in selecting the topics, it is toward more recently developed methods. For example, the familiar comparative techniques related to the linear model are only selectively covered; those that are included are the most representative and widely used. On the other hand, generalized linear, latent variable, and multilevel models are discussed in greater detail. Some comparative methods included in the book, such as the double generalized linear model and the relative distribution methods, were only developed and refined in the last few years and require use of special software. My hope is to bring these methods to a broader cadre of applied researchers, because I believe they are of great potential for our comparative research.

Comparative methods have always been my research interest. It was not until 1999, however, that I formed the idea of this book while on sabbatical at the University of Cambridge from the University of Illinois. Perhaps it was this trans-Atlantic residence that heightened my awareness of the importance of making comparisons. But it was the warm reaction to my prospectus from and the forward-looking vision of Stephen Quigley, Senior Editor at Wiley, that really got me going. Eight of the chapters were written in the fall of 2000 while teaching a graduate applied statistics course on comparison at the University of Illinois, and I finished the final draft of the manuscript at the University of Essex.

I would like to express my gratitude to the people who have facilitated or helped the writing of the book: to Ken Bollen and Noel Cressie, who gave constructive feedback to the prospectus; to Scott Long, who read a draft of seven of the chapters and gave invaluable comments; to all the students in my statistical comparison seminar, who served as guinea pigs in the class where I used the manuscript as the main text and gave me feedback, and especially to Ringo Moon-Ho Ho, who caught typos, inaccuracies, and errors of various kinds; to Tom Snijders for use of the Dutch schools data in Sections 10.3 and 10.4; to Jim Kluegel for use of the international social justice in data in Section 10.4; to Mark Handcock for introducing me to the relative distribution methods; to Graham Upton for catching a number of mathematical mistakes in the page proofs. The remaining errors, which are bound to be there with a project of this coverage, are all mine.

I would like to thank Stephen Quigley, Heather Haselkorn, and Rosalyn Farkas of Wiley Publications and Joseph Fineman for their understanding and assistance in this project. Finally, my gratitude goes to my family, Keiko and Athena, whose love and support have kept me focused on this project, and to whom this book is dedicated.

TIM F. LIAO

Colchester, England
January 2002

CHAPTER 1

Introduction

1.1 RATIONALE FOR STATISTICAL COMPARISON

As a boy of eleven I read a book about an imaginary world without friction where no one and nothing could stay still anywhere. Now imagine a world without *comparison*. The consequences would be no less severe if somehow we were deprived of the ability to compare. We would read any book that came our way, buy any grocery that stocked the supermarket, and worst of all, get true lemons when it came to purchasing a used automobile! We would not have any competitive sports, for it would be senseless to compete because "faster," "higher," or "farther" would not necessarily mean better or more desirable.

Such a world is unthinkable. Indeed, comparison is an indispensable part of our lives. As Joan Higgins put it, "acts of comparing are part of our daily lives," as parents compare their babies "to see whether they are unusually fat or thin, or small or large" (Higgins 1981, p. 7). No one would likely disagree that making comparison is one of the most important activities in human thinking process. Without such activity, the world would be a truly boring place in which to live. The process of comparison is mundane, yet indispensable in our daily lives.

Everybody's ability to distinguish notwithstanding, great philosophers still wrote about the import of comparison, simply because human knowledge cannot exist without it. Descartes, for example, once commented, "It is only by way of comparison that we know the truth precisely.... All knowledge which is not obtained through the simple and pure intuition of an isolated thing is obtained by the comparison of two or more things among themselves. And almost all the work of human reason consists without doubt in making this operation possible" (Descartes 1963, p. 168; translation by Lacour 1995). Therefore, according to Descartes, knowledge can be obtained through two means, either by simple, pure, intuitional observation of isolated things, or by comparison, when the isolated thing is not the object of knowledge, but theoretical reflection, as opposed to pure and simple understanding, leads to knowledge. For Descartes, the only thing we may observe directly and intuitively is spatial extension; for all other things, including the thinking thing that is the self, we must resort to the method of comparison. The fundamentally nonintuitional things that are cultural, social, or otherwise abstract become our knowledge not in themselves but by way of comparison. For Descartes, the basis for reason is comparison.

The comparative domain is not confined to things cultural and social. Another great thinker, Goethe, devoted much of his lifetime to the composition of a "comparative morphology" in the fields of anatomy, botany, geology, mineralogy, and osteology (Lacour 1995). He stated the centrality of comparison to the study of natural phenomena in the following words: "Natural history rests on comparison. External characteristics are significant, but not sufficient for the task of properly separating organic bodies and joining them back together again" (Goethe 1982, p. 170; translation by Lacour 1995). Goethe was a practitioner of scientific comparison. The principle of comparison that could reveal the underlying continuity guided his celebrated identification of the missing link in the natural history relating apes to humans. His comparative study of ape and human osteology demonstrated the shared traits of animal and human anatomy and hence common genealogy. For Goethe, anyone who looks and compares can make discoveries of this order, and comparison is the essential complement to empirical vision (Lacour 1995). Goethe made the discovery by comparing visible characteristics to reveal what was previously imperceptible in isolation. Yet Goethe applied the comparative method to not only things alike, but also things unalike, to gain knowledge and find underlying principles of nature.

Just like Goethe, Descartes made tremendous contributions to scientific knowledge through the comparative method. To facilitate comparison, he put forth analytic means such as Cartesian coordinates and polynomial equations. He used concise signs to represent things to compare in the empirical world. It is the same analytic methods of Descartes that serve as the foundation for making statistical comparisons today.

This book is about how to make statistical comparisons in applied research. We compare things both for their similarities and for their differences. Comparison is of course never just finding similarities, identities, or equalities, nor is it just finding dissimilarities, differences, or inequalities. Some view comparison as a dialectic process of finding differences between things that appear to be similar and of finding similarities between things that appear to be different (Mitchell 1996). In that respect, a "compare and contrast" essay assignment to a class makes explicit by the word "contrast" what is already implicit in the notion of comparison. Instead, I would rather consider similarities and differences along a continuum. That is, an act of comparison produces an entire spectrum of results ranging from outright identity to striking difference. In this way, we may perform statistical tests of comparison that either retain the hypothesis of equality or support the hypothesis of inequality with a varying degree of evidence.

Furthermore, the methods discussed in the book may apply to both cross-sectional and longitudinal situations. In other words, we treat difference between groups at the same time and between different time points in an identical manner. Time generates difference, as does a change of groups. Do these differences differ? According to the noted American philosopher John William Miller (1981), the answer is a resounding yes. For him, there exists a fundamental distinction between history and physics with regard to how time and difference are viewed in each discipline. In history, change in time generates difference, and all difference is original and thereby ontological. In physics, the goal is to explain how one event is different from another, and here difference is derivative rather than ontological. For our purpose of making statistical comparison, differences over time and between things at the same time are both derivative; yet such derivatives help contribute to original understanding of fundamental social and natural mechanisms. The new knowledge, hopefully, is ontological.

1.2 COMPARATIVE RESEARCH IN THE SOCIAL SCIENCES

A growing practice in the social sciences is the comparison of human behavior between social groups. The behavior in focus can be economic, political, psychological, or social at the individual or the organizational level, and the groups of interest often are racial, ethnic, gender, or national ones. Recent social trends in understanding and emphasizing multiculturalism, inclusivity, and globalization may further stimulate comparative research in the twenty-first century. Although the examples cited below are confined to the social sciences, statistical comparison is commonly conducted in other disciplines, especially the biological and medical sciences. The discussion below only demonstrates the tip of one iceberg, let alone other icebergs. The generality and importance of comparative research cannot be overestimated.

Comparative research has its roots in the beginning of the social sciences. For many classical social writers, who studied and compared all societies for which they could find any data, the notion of "comparative" study would sound redundant, for social science had to be comparative by definition (Nowak 1989). Indeed, it has been widely recognized that Weber's social theories provide a solid basis for cross-national comparison (Wesolowski 1989). Today's emphasis on comparative research, as evidenced by the selected references given below, only represents a resurgence of interest in social comparison.

Making substantively informative statistical comparisons has become an important analytic strategy for social scientists from a variety of disciplines. Researchers from anthropology, economics, psychology, political science, and sociology have always been interested in making systematic comparisons between human groups, institutions, organizations, or societies. Viewed from the perspective of empirical research, most comparisons fall into two large categories—comparisons of gender, ethnic, and racial groups, and cross-cultural and cross-national comparisons.

Examples of either type abound in the social science literature. Many recent studies focus on gender differences in economic, political, psychological, or social behavior (e.g., Arber and Cooper 1999; Echevarria and Merlo 1999; Henriques and Calhoun 1999; Johnson and Marini 1998; McDowell, Singell, and Ziliak 1999; Perrucci, Perrucci, and Targ 1997; Sapiro and Conover 1997). In these same social science disciplines we may also find research on racial and ethnic differences in a range of human behavior (e.g., Choi 1997; Henriques and Calhoun 1999; Lloyd and South 1996; Mennen 1994; Telles and Lim 1998). The references above reflect merely the tip of a huge iceberg of empirical social science research that examine racial, ethnic, and gender differences.

The trend of cross-national, cross-cultural has gained greater momentum in the last decade. Again, many studies belonging in this tradition are difficult to miss, appearing in top journals in the social sciences (e.g., Alderson and Nielsen 1999; Bornstein, Haynes, and Azuma 1998; Browne 1998; Ehrhardt-Martinez 1998; Ellina and Moore 1990; Hupkens, Knibbe, and Otterloo 1998; Ishida, Muller, and Ridge 1995; Kenworthy 1999; Sullivan and Transue 1999; Westholm and Niemi 1992; Wright, Baxter, and Birkelund 1995). While the economic, social, and behavioral sciences have abundant examples of cross-national research, educational studies that compare a number of societies have also been frequent in the literature (e.g., Becker 1992; Cai 1995; Silver, Leung, and Cai 1995).

The empirical social research using data from multiple nations can be distinguished by the intent of the research as one of four types: nation as object of study, nation as context of study, nation as unit of analysis, and nation in transnational setting (Kohn 1989). Despite the criticism on the validity of cross-national research by researchers of the transnational persuasion in sociology (e.g., McMichael 1990; Martin and Beittel 1998), cross-national or comparative research enjoys ever greater popularity in the broader social sciences, evidenced by the recent examples referred to earlier. This is also demonstrated by the particular attention paid to the comparative method in individual social science disciplines, such as the discussions in *Current Anthropology* a few years ago and the new book edited by Arts and Halman (1999) on recent directions in quantitative comparative sociology.

However, there has not been a systematic treatment of the quantitative methodology for making valid group comparison. Social science researchers tend to ignore any formal testing when making social comparison (as in many of the references above). One explanation might be that making comparison is not typically a primary focus when researchers learn a particular statistical model, and it is also true that most applied statistics texts barely cover the topic (with the exception of comparing group means and proportions), much less do so in a systematic way.

In this book we are concerned about statistical comparison between groups. For the most part our focus is on parametric methods and models, though we will include some nonparametric methods as well. The major merit of a parametric method is that the same test can be carried out over many different models as long as the assumptions of the parametric method are still valid. Furthermore, nonparametric tests are more developed for simpler analyses than for more advanced statistical models. Similarly, we will focus our attention on the frequentist approach in statistics, because it is more developed for comparative purposes; we treat the Bayesian approach as an extension.

1.3 FOCUS OF THE BOOK

Comparative studies are an important way of conducting research in many a discipline. A Wiley book by Sharon Anderson and colleagues (1980) impressively summarized many statistical methods for comparative purposes. The methods examined include standardization, matching, analysis of variance, logit and loglinear analysis, and survival methods. The book has over the years facilitated researchers in carrying out their comparative research while dealing with the problems of confounding variables.

However, the focus of the book is precisely on confounding factors in comparison, as indicated by the title, "Statistical Methods for Comparative Studies: Techniques for Bias Reduction." The current book, in contrast, assumes that the researcher is aware of potential confounding factors, and has included them in the statistical model of concern, be it a linear, a generalized linear, or a generalized linear mixed model. The only exception is rate comparison, which conventionally is not modeled with the effects of confounding factors captured by parameters. Thus, we consider rate comparison separately.

Indeed, statistical developments and their applications in a variety of biomedical and social sciences have popularized not only the knowledge of the potential consequences of confounding variables but also the common ways of dealing with them. The biases due to certain confounding factors can be represented by parameters in the model, and their magnitudes across multiple groups are of interest as well. These parameters, together with others representing the effects of other variables of interest, constitute the parameter vector for comparison. Therefore, we focus in this book on the actual comparative methods themselves.

1.4 OUTLINE OF THE BOOK

Following this introduction, the book examines statistical comparison between groups in nine substantive chapters. Groups for many considerations are regarded as fixed, though we also study random groups in multilevel analysis. In the social and biological sciences, viewing groups as fixed most often is a reasonable approach and is a common practice. In Chapter 10 we will consider groups from a random distribution.

1.4.1 Chapter 2—Statistical Foundation for Comparison

This chapter introduces the systematic approach for making statistical group comparisons in many statistical models. We then consider six usual test statistics: the t, the χ^2, the F, the likelihood ratio, the Wald, and the Lagrange multiplier. The last three are asymptotic equivalents.

1.4.2 Chapter 3—Comparison in Linear Regression

The chapter begins with the simplest comparison, comparing means and proportions between groups. This is then viewed as a special case of testing parameter equality in linear regression. We also consider some multiple comparison techniques: analysis of variance, analysis of covariance, and regression decomposition.

1.4.3 Chapter 4—Nonparametric Comparison

In this chapter we consider several major methods of nonparametric comparison. These include nonparametric tests, as well as resampling methods which can be applied to otherwise parametric tests. We also discuss the recent development of relative distribution methods.

1.4.4 Chapter 5—Comparing Rates

Social and biomedical researchers use rates for making comparisons as well as assessing the intensity of event occurrence. Often compositional structure may confound the comparison. In this chapter we present the methods of standardization and decomposition of both the conventional arithmetic type and the model-based approach.

1.4.5 Chapter 6—Comparison in Generalized Linear Models

The chapter first introduces the family of models known as generalized linear models; it then examines group-based parameter comparisons in such models as logit, probit, multinomial logit, and hazard rate models, which are all members of the generalized linear model. This chapter is important in that the generalized linear model provides a framework for linking previously unrelated methods.

1.4.6 Chapter 7—Additional Topics of Comparison in Generalized Linear Models

We consider three extensions: matched studies, dispersion heterogeneity, and Bayesian modeling. While the Bayesian version of generalized linear models can be viewed as an extended use of these models and heterogeneity can be seen as a nonstandard condition in the generalized linear model, matched studies are a special case when generalized linear models are used in designs matching treatment and control groups on certain characteristics. Such designs are useful not only in disciplines where experiments are carried out, but also in disciplines where quasiexperiments are employed. Thus, match studies deserve our special attention as a special type of statistical group comparison.

Sometimes group comparison in generalized linear models may be affected by nonstandard conditions, especially when dispersion is heterogeneous. Heteroscedasticity in linear regression is just a special case, and dispersion heterogeneity may occur in other generalized linear models. In the chapter we study a logit model with dispersion heterogeneity, and employ the quasi-likelihood-ratio test for making group comparisons. Next, we present a version of Bayesian statistics, generalized linear Bayesian modeling, as an extension of generalized linear models. When making group comparison in generalized linear Bayesian models, we compare not just two competing models of group identity versus group difference; instead, we allow for all possible competing models in between the two extremes in the model space when testing the equality hypothesis.

1.4.7 Chapter 8—Comparison in Structural Equation Modeling

So far the models considered have been based on a single equation. In this chapter we consider multiequation systems in which the exogenous and endogenous variables can be either observed or latent. In such system group comparison is also desirable. Group differences in structural equation modeling may take on different forms: As with the models considered so far, the hypothesized parameters may be different; furthermore, the hypothesized structures of the model may be different, or both conditions may be true. In addition, the comparison may involve covariance structure, mean structures, or both. In comparison with computer programs for comparing generalized linear models, the syntax for comparing structural equation models can be much more complex. Therefore, EQS and LISREL program listings for the examples are supplied in an appendix.

1.4.8 Chapter 9—Comparison with Categorical Latent Variables

In structural equation models latent variables are assumed continuous. However, latent variables can also be conceptualized and treated as discrete.

This chapter considers various types of models of categorical latent variables but concentrates on group comparison in latent class analysis and its extension by including latent (class) variables in conventional path models. The latent class model can also be parameterized as a loglinear model, and such parameterization makes it possible to integrate all major developments in loglinear modeling into latent class analysis. We discuss group comparison in these variants of latent class modeling. Because there are many special-purpose software packages available for this type of models, an appendix lists the major ones. Another appendix gives the computer program listings for the examples.

1.4.9 Chapter 10—Comparison in Multilevel Analysis

In recent years multilevel analysis has become increasingly popular in medical, educational, and social science research. How do we handle group comparisons in such analysis? In this chapter we consider treating group as one of the levels in the analysis and contrast such treatment with testing equality in generalized linear models previously discussed. Group can be considered as coming from a random distribution, instead of as fixed units, in this approach.

1.4.10 Summary

While there is a general increase in complexity in the methods discussed over the chapters, researchers familiar with a particular method yet interested in learning how to make comparison using the method may go to the relevant chapter directly without much trouble. For example, those who have been exposed to structural equation modeling can consult the topics in Chapter 8, possibly after a quick browse of the basics in Chapter 2.

CHAPTER 2

Statistical Foundation for Comparison

2.1 A SYSTEM FOR STATISTICAL COMPARISON

In order to facilitate statistical comparison with a wide array of methods and models, we define below a simple, generic system of testing equality versus inequality. The system applies to statistical group comparison from the very basic, such as comparison of means, to the quite advanced, such as latent variable models.

Let us begin by introducing the typical null hypothesis:

$$H_0: \quad \mathbf{S}_1 = \mathbf{S}_2 = \cdots = \mathbf{S}_G.$$

The statistic $\mathbf{S}$ is a vector of particular statistics or parameters of interest, which can be means, proportions, regression coefficients, and so on. The subscript G gives the total number of groups to be compared. The test statistic for the hypothesis is a function taking the form

$$d(\mathbf{S}_g) \quad \text{for } g = 1, 2, \ldots, G.$$

The function $d(\cdot)$ gives rise to the t, F, and χ^2 test statistics based on their respective distributions with cutoff values corresponding to specific degrees of freedom. The goal of statistical comparison, regardless of $d(\cdot)$, is to show whether the null hypothesis is to be retained or rejected. Although most often this function follows a parametric distribution, it may also be defined to represent nonparametric tests and others. For example, the function can also be that of a permutation test.

Following the generality of various parametric models, in this chapter we focus our attention first on the t, F, and χ^2 test statistics. It will become clear that statistical group comparison in many parametric methods follow the same rationale and use the same statistical tests. The issues of nonparametric and permutation tests will be taken up in subsequent chapters whenever applicable.

2.2 TEST STATISTICS

The following tests are the so-called significance tests. In them, random sample data are assumed. As with all significance tests, if you have the whole population, then any differences are real and therefore significant. If you have nonrandom sample data, statistical significance cannot be established, though significance tests are nonetheless employed sometimes to obtain a crude "rule of thumb" sense of strength (of difference). It is important to note that the tests described below are all based on their respective distributions. However, these distributions do not constitute tests themselves, but describe the behavior of a test. For each test, we introduce first the underlying distribution and then the test.

2.2.1 The χ^2 Test

Suppose that we have a population of cases with variable Y following the normal distribution. Now we sample the cases from this population one at a time, $N = 1$. For each sample, we first standardize Y by computing its score Z, then square it to obtain Z^2. Let us call this squared standardized score $\chi^2_{(1)}$, so that

$$\chi^2_{(1)} = Z^2. \tag{2.1}$$

Thus, we can generate a χ^2 distribution from a population of standard normal values. Equation (2.1) gives a chi-square distribution with one degree of freedom, with its well-known curve sloping sharply to the right. Because $\chi^2_{(1)}$ is a squared quantity, its values must be all nonnegative real numbers, ranging from zero to positive infinity.

Now suppose that two cases are drawn independently and randomly from the earlier population. Then the sum of the two squared standardized scores is found over repeated samplings, with the resulting random variable as $\chi^2_{(2)}$:

$$\chi^2_{(2)} = Z_1^2 + Z_2^2. \tag{2.2}$$

This is a chi-square distribution with two degrees of freedom. The distribution of $\chi^2_{(2)}$ is a little less skewed than that of $\chi^2_{(1)}$.

We further suppose that N independent random observations are drawn from the previously described normal distribution, resulting in the random variable

$$\chi^2_{(N)} = \sum_i Z_i^2 \quad \text{for } i = 1, 2, \ldots, N. \tag{2.3}$$

The distribution of $\chi^2_{(N)}$ has a shape that depends on the number of independent cases drawn randomly at a time. For N independent cases drawn randomly from a normally distributed population, the sum of the squared standardized scores of the cases follows a chi-square distribution with N degrees of freedom.

Because $Z_i = (y_i - \mu)/\sigma$, then

$$\sum_{i=1}^{N} \left(\frac{y_i - \mu}{\sigma} \right)^2 = \frac{\sum_{i=1}^{N} (y_i - \mu)^2}{\sigma^2} \sim \chi^2_{(N)}.$$

We generalize the result and express it in matrix notation:

$$\frac{1}{\sigma^2}(\mathbf{y} - \boldsymbol{\mu})'(\mathbf{y} - \boldsymbol{\mu}) = (\mathbf{y} - \boldsymbol{\mu})'(\sigma^2 \mathbf{I})^{-1}(\mathbf{y} - \boldsymbol{\mu}) \sim \chi^2(N). \tag{2.4}$$

The middle factor $\sigma^2 \mathbf{I}$ is the covariance matrix of y_i, which can be replaced by a more general $\sigma^2 \mathbf{W}$. This will prove a useful form for tests such as the Wald in later sections.

As with Student's t-distribution to be discussed later, there is no longer just one distribution, but a distribution for each specific degree of freedom. The distribution curves are strongly skewed to the right at lower degrees of freedom, but more or less approach symmetry for higher degrees of freedom. This phenomenon can be understood as a result of where the bulk of cases are located in a normal distribution. For example, about 68% in the standard normal distribution must lie between -1 and 1. This means the corresponding $\chi^2_{(1)}$ value must lie between 0 and 1, thus giving rise to the very skewed shape. With summation over two or more independently drawn random observations, the corresponding $\chi^2_{(2)}$ or in general $\chi^2_{(N)}$ must have the bulk of cases shifted away from zero to greater values, thereby showing a more symmetric distribution. As a visual aid, the chi-square sampling distribution generator located at http://vassun.vassar.edu/~lowry/csqsamp.html on the Internet is helpful not only in understanding the relation between the form of the distribution and the degrees of freedom but also in getting a sense of associated critical values at the various common levels from 0.05 to 0.001.

There are many significance tests relying on the chi-square distribution. The best known is probably Pearson's chi-square, which is widely used in contingency table analysis. The likelihood ratio, the Wald, and the Lagrange multiplier test statistics, to be discussed later, all follow the chi-square distribution.

2.2.2 The *t*-Test

Let $\chi^2_{(k)}$ be a chi-square variable whose number of degrees of freedom is df = k. The *modified* $\chi^2_{(k)}$ is obtained by dividing by k (Wonnacott and Wonnacott 1979):

$$C_k^2 \equiv \frac{\chi^2_{(k)}}{k}.$$

Because $\chi^2_{(k)}$ may be considered as a sample sum, C_k^2 may be considered as a sample mean from the Z_k population with 1 as its expectation and $2/k$ as its variance. The mathematical definition of Student's t with k df is defined as

$$t_k \equiv \frac{Z}{C_k}. \qquad (2.5)$$

This *t*-distribution is known as Student's distribution in honor of W. G. Gossett, who first published it under the pseudonym "Student" in the early 1900s. His employer then had regulations that prevented him from publishing his discovery under his real name.

The *t*-distribution shares with the normal certain characteristics. Both are symmetric and extend from negative to positive infinity. But the *t*-distribution differs from the normal in that it assumes shapes depending on the number of degrees of freedom. In contrast, the Z-ratio, which is not dependent on sample size or degrees of freedom, follows the standard normal curve. Because the number of degrees of freedom can range from 1 to infinity, the shape of the *t*-distribution differs from that of the normal. A *t*-distribution for df = 1 deviates the most from the standard normal, and Student's distribution approaches the shape of the standard normal distribution as the number of degrees of freedom increases. For df = 30 it becomes less indistinguishable from the standard normal. At df = ∞, the two distributions are identical. In this sense, we can think of the normal as a special case of Student's distribution with df = ∞.

The *t*-statistic is closely related to the Z-statistic, which is the ratio of a deviation statistic, a random variable in itself, to the standard error of the sampling distribution of that statistic. The Z-ratio follows a standard normal distribution. The Z-ratio assumes a known population variance and hence a known standard error of the statistic. However, in most situations we have only sample data, and we do not know the true values of the population variances. Therefore, we estimate the standard error of a statistic using its sample variance. The ratio of a statistic to its estimated standard error forms a *t*-statistic. The *t*-test is based on the *t*-distribution, which is the expected distribution of the *t*-ratio.

A common use of the *t*-statistic is for testing the difference between two group means from groups g and h takes the form

$$t_{\text{df}} = \frac{(\mu_g - \mu_h) - 0}{\hat{\sigma}_{\mu_g - \mu_h}} = \frac{\mu_g - \mu_h}{\hat{\sigma}_{\mu_g - \mu_h}}. \qquad (2.6)$$

TEST STATISTICS

Here we assume that the null hypothesis states the difference to be zero, and thus omit the zero difference from the numerator.

When we substitute the sample variance $\hat{\sigma}$ for σ, we obtain a quasistandardized variable. In general, whenever any normal variable is quasistandardized, it becomes a t-variable rather than a Z-variable. The regression coefficient $\hat{\beta}$ is a good example. The quasistandardized variable is a random variable, and the confidence intervals as a result will be a little wider than the ones related to Z.

2.2.3 The F-Test

The F-distribution, or the distribution of the random variable F, is named after Sir Ronald Fisher, who developed the concept as well as some major applications of this distribution. The mathematical definition of an F-variable with df_1 and df_2 is similar to that of t:

$$F_{[df_1, df_2]} \equiv \frac{C^2_{df_1}}{C^2_{df_2}} \tag{2.7}$$

where $C^2_{df_1}$ and $C^2_{df_2}$ are independent modified chi-square variables. When they are perfectly dependent, the ratio becomes 1, which is no longer a random variable.

Another common use of the F is for testing variance equality. Often we need to compare two population variances, and the F-distribution is the distribution of the ratio of their estimated variances. Imagine that we have two distinct populations, each following a normal distribution of the variable Y. We draw two independent samples from the two populations, the mean of which may be different, but the variances of which are the same, indicated by σ^2. The two random samples have the sizes of N_1 and N_2, respectively. For each pair of samples from the two populations, we take the ratio of the estimated variances $\hat{\sigma}^2$, or S_1^2 to S_2^2, which follows the F-distribution:

$$F = \frac{S_1^2}{S_2^2}. \tag{2.8}$$

Notice that by taking the ratio of the two variance estimates, given that their hypothesized population variances are equal, we actually specify the ratio of two independent chi-square variables, each divided by its own number of degrees of freedom:

$$F_{[df_1, df_2]} = \frac{\chi^2_{(df_1)}/df_1}{\chi^2_{(df_2)}/df_2}. \tag{2.9}$$

Such a random variable is called an F-ratio, and follows the F-distribution. Regardless of the mathematical complication of the F-ratio computation, it suffices to remember that the density of F depends only on two parameters, df_1 and df_2, the numbers of degrees of freedom associated with the numerator and the denominator of the ratio. The shape of the distribution is a function of both numbers of degrees of freedom. To appreciate the relation between the degrees of freedom and the F density function, the Web site http://vassun.vassar.edu/~lowry/fsamp.html may be helpful. As with the chi-square sampling distribution generator, the user supplies degrees of freedom to generate a corresponding F density function.

There exist many applications for the F-test, especially in analysis of variance and linear regression. The F-statistic for testing a simultaneous equality hypothesis in a linear multiple regression with k parameters takes the form

$$F_{[df_1, df_2]} = \frac{\text{SSM}/(k-1)}{\text{SSE}/(N-k)}.$$

That is, the ratio of the sum of squares of the model to the sum of squared errors, divided by their associated degrees of freedom, respectively, forms an F-ratio.

2.2.4 The Likelihood Ratio Test

Next we consider the likelihood ratio test. Given a sample of observed data of $\{n_i, i = 1, 2, \ldots, N\}$, the likelihood function, L, is the probability of jointly observing $\{n_i\}$ as a function of some unknown parameters $\boldsymbol{\theta}$. L is defined by the particular model the researcher uses to analyze the data, and takes a particular form according to the underlying distribution of the model. The maximum likelihood (ML) estimation is the procedure to obtain estimates as parameter values that maximize the likelihood function L. Suppose our null hypothesis is $\boldsymbol{\theta}_0 = 0$. The null hypothesis can be viewed as a restriction applied to all parameters. More generally we may express the vector $\boldsymbol{\theta}$ of parameters as $\mathbf{S}$ for any statistics to be tested, with the null hypothesis $\mathbf{s}_1 = \mathbf{s}_0$ and the alternative hypothesis $\mathbf{s}_1 \neq \mathbf{s}_0$. Now let L_1 be the maximum value of the likelihood of the data without the additional restriction. In other words, L_1 is the likelihood of the data with $\boldsymbol{\theta}$ unrestricted and ML estimates substituting for it, denoted $\boldsymbol{\theta}_1$. Let L_0 be the maximum value of the likelihood for the restricted vector $\boldsymbol{\theta}_0$ based on the null hypothesis. The dimension of $\boldsymbol{\theta}_0$ is $k \times 1$ (i.e., L_0 has k less parameters than L_1).

Next, we form the ratio L_0/L_1. The value of this ratio always falls between 0 and 1, and the less likely the null hypothesis is to be true, the smaller the ratio will be. This ratio is the basis for the likelihood ratio test (LRT) statistic. In practice, we calculate the LRT at a given confidence level by forming $-2\ln(L_0/L_1)$. In other words, we multiply the natural logarithm of the likelihood ratio by -2. Alternatively, we obtain the LRT statistic by taking the difference of -2 times the log-likelihood values, or the difference

TEST STATISTICS

of $-2\ln L_1$ and $-2\ln L_0$:

$$\text{LR} = -2(\ln L_1 - \ln L_0). \quad (2.10)$$

Because the LRT statistic follows a chi-square distribution, we can check the significance level of the test by comparing the LRT statistic with the upper $100 \times (1 - \alpha)$ percentile point of a chi-square distribution with k degrees of freedom. The approximation of the distribution of LRT statistics to the chi-square distribution is usually good, even for small sample sizes. The LRT computed accordingly rejects the null hypothesis if it is greater than the critical value from the chi-square distribution with k-df freedom percentile. The analyst chooses the percentile, which corresponds to the confidence level of the test.

Let us return to the general situation of assessing **S** when there is only one element in it and describe the LRT as shown in Figure 2.1. Suppose that we have a ML estimation of a certain statistic of interest, $\mathbf{s}_1$, whose log-likelihood function represented by curve a. Now if the restricted value of **S** is $\mathbf{s}_0$, then the LRT statistic is given by the difference of $\ln L_1$ and $\ln L_0$, which is chi-square distributed. The distance between $\ln L_1$ and $\ln L_0$ is given by half the likelihood ratio. The curvature of the log-likelihood function a is denoted by $C(\theta_1)$ and defined by the absolute value of $d^2\ln L/d\theta^2$ evaluated at $\theta = \theta_1$ (Buse 1982). Given a certain distance between $\mathbf{s}_1$ and $\mathbf{s}_0$, the greater the curvature $C(\theta_1)$, the larger the distance (half the likelihood ratio) will be. Conversely, given a certain curvature $C(\theta_1)$, the larger the distance between $\mathbf{s}_1$ and $\mathbf{s}_0$, the further away $\ln L_0$ will be from the maximum $\ln L_1$. This diagrammatic illustration will also assist the understanding of the Wald test.

2.2.5 The Wald Test

Unlike the LRT, which is characterized by the distance between $\ln L_0$ and the maximum of $\ln L_1$, the Wald test, due to Wald (1943), considers the squared distance between $\mathbf{s}_1$ and $\mathbf{s}_0$. The greater the squared distance, the stronger the evidence for rejecting the null hypothesis will be. However, because the curvature affects the behavior of this test just as it does the LRT as discussed in the previous section, the squared distance between $\mathbf{s}_1$ and $\mathbf{s}_0$ must be weighted by the curvature $C(\theta_1)$. Two sets of data may generate two curves with different curvatures but produce identical distances $\mathbf{s}_1 - \mathbf{s}_0$, with one set more favorable to the hypothesis than the other if we examine the LRT. Another look at Figure 2.1 will help illustrate the point. The curves a and b arise from two different sets of data, yet the $(\mathbf{s}_1 - \mathbf{s}_0)^2$ values are the same. However, curve a gives a greater value of $\ln L_0$ than curve b, whose $\ln L_0$ is indicated by $\ln L_b$. Thus, a Wald statistic is a weighted squared distance such as

$$W = (\mathbf{s}_1 - \mathbf{s}_0)^2 C(\mathbf{s}_1 - \mathbf{s}_0),$$

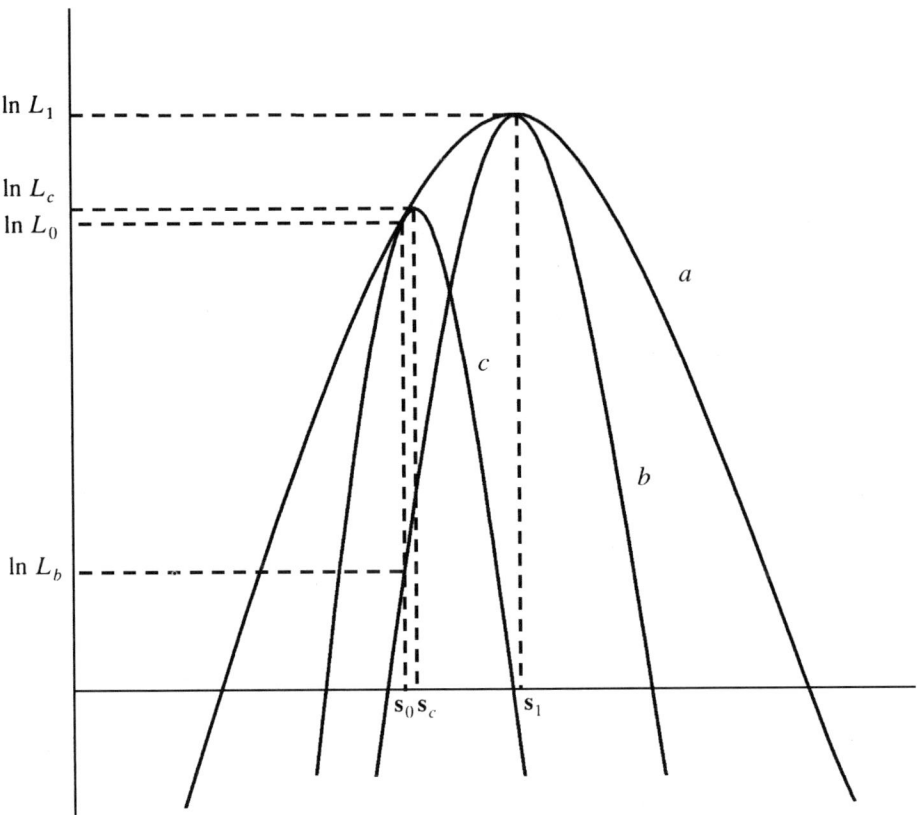

Figure 2.1 A comparison of the three asymptotic tests.

which is asymptotically distributed as χ^2 with one degree of freedom. The formula can be replaced by the more common form,

$$W = (\mathbf{s}_1 - \mathbf{s}_0)^2 I(\mathbf{s})(\mathbf{s}_1 - \mathbf{s}_0) \tag{2.11}$$

where

$$I(\mathbf{s}) = \frac{d^2 \ln L}{d\mathbf{s}^2},$$

known as the information matrix. Now the squared distance is weighted by the average curvature.

As a widely used alternative to the LRT in hypothesis testing, the Wald test (WT) also offers a more straightforward analogy to the t-test, at least in the single parameter case. The test takes the following general form for testing $H_0: \boldsymbol{\theta} = 0$:

$$W = \hat{\boldsymbol{\theta}}_1' \big[\text{cov}(\hat{\boldsymbol{\theta}}_1) \big]^{-1} \hat{\boldsymbol{\theta}}_1, \tag{2.12}$$

where $\hat{\boldsymbol{\theta}}_1$ represents estimated values of $\boldsymbol{\theta}$ under the unrestricted model. When $\boldsymbol{\theta}$ has a single element (i.e., when df = 1), W is simply the square of the ratio of the parameter estimate to its estimated standard error, and is reported for many logistic regression procedures in major software packages.

The Wald statistic is an asymptotic equivalent to the LR statistic, and approximates the chi-square distribution as well. The Wald principle is based on the idea that if a restriction on parameters is true, the unrestricted model should approximately satisfy the restriction. The WT statistic, which can be fairly simple to compute even if not given by a computer output, shows how much the unrestricted model deviates from the restricted ideal.

2.2.6 The Lagrange Multiplier Test

The Lagrange multiplier test (LMT), also known as the score test, is due to R. A. Fisher and C. R. Rao and, like the WT, is another alternative to the LRT for testing a hypothesis such as $\boldsymbol{\theta} = \boldsymbol{\theta}_0$. The discussion of curvatures of a log-likelihood function also applies to the LMT, but the test pertains to the properties of the function when restrictions are imposed. If the null hypothesis is true, the restricted $\ln L_0$ will be close to the unrestricted $\ln L_1$. Because the unrestricted estimates maximize the log likelihood, they satisfy the condition $d(\mathbf{s}_1) = 0$, where $d(\mathbf{s}_1) = d \ln L / d\mathbf{s}_1$. The squared score, or $[d(\mathbf{s}_1)]^2$, is suggested as a test statistic because the sign of the slope of the function is irrelevant.

We note that two different data sets may give rise to two different log-likelihood functions with identical slopes, with one curve producing an maximum $\mathbf{s}_1$ that is closer to $\mathbf{s}_0$. Figure 2.1 illustrates the point. Suppose that another set of data is responsible for the log-likelihood function represented by curve c. The new $\mathbf{s}_1$ should correspond to the position of the maximum of curve c on the x-axis, designated as $\mathbf{s}_c$. The difference $\mathbf{s}_c - \mathbf{s}_0$ and the difference $\ln L_c$ and $\ln L_0$ are both smaller. The slopes for the log-likelihood functions represented by curves a and c at the restricted value of $\mathbf{s}_0$ (where the two curves touch) are identical. This phenomenon is due to the difference in curvature of the likelihood functions. Once again, to correct this problem we weight the squared slope by the curvature. Nevertheless, in this case, the greater the curvature, the closer $\mathbf{s}_0$ will be to the maximum of the log likelihood. That suggests weighting by the inverse $C(\mathbf{s}_0)^{-1}$ to ensure smaller values of the test statistic relating to a smaller distance between $\mathbf{s}_0$ and $\mathbf{s}_c$:

$$\text{LM} = [d(\mathbf{s}_0)]^2 C(\mathbf{s}_0)^{-1},$$

which, as before, can be replaced by the more common form

$$\text{LM} = [d(\mathbf{s}_0)]^2 I(\mathbf{s}_0)^{-1}, \qquad (2.13)$$

which follows asymptotically a χ^2 distribution with one degree of freedom.

More generally, the test can be viewed as a quadratic function based on the vector of partial derivatives of ln L with respect to θ, evaluated at $\hat{\theta}$ assuming H_0. Mathematically, it takes the form

$$\text{LM} = \left[d(\hat{\theta}_0)\right]' \left[I(\hat{\theta}_0)\right]^{-1} \left[d(\hat{\theta}_0)\right], \tag{2.14}$$

where $d(\theta_0)$ contains the derivatives of the log-likelihood function evaluated at the restricted values, $\partial \ln L(\hat{\theta}_0)/\partial \hat{\theta}_0$ contains the derivatives of log-likelihood function evaluated at the restricted values, and $I(\hat{\theta}_0)$ is the information matrix. The LMT is asymptotically equivalent to the LRT and WT, though it is evaluated at the restricted values only.

2.2.7 A Summary Comparison of the LRT, WT, and LMT

A practical shortcoming of the LRT is that it requires estimation of both the restricted and unrestricted models, though that is much less a problem for today's fast computers than it used to be. In complex models, one or the other of these estimates may be very difficult to compute. It would be unfortunate if this precluded testing of the hypothesis. Fortunately, the WT and LMT circumvent the problem. In a typical scenario of testing parameter significance, the WT relies only on the unrestricted estimator, and the LMT only on the restricted estimator. Both alternative tests are based on an estimator that is asymptotically normally distributed, and on the distribution of the full rank quadratic form. All three are asymptotically equivalent, while their small-sample properties are largely unknown and can behave rather differently in small samples (Greene 1990). For large samples they all approximate the chi-square distribution.

The three asymptotically equivalent tests also are based on quite different theoretical principles related to the likelihood function. A revisit to Figure 2.1 reveals how LRT, WT, and LMT compare. For testing s_1 against s_0, the LRT statistic is given by the difference of ln L_1 and ln L_0, which is chi-square distributed. Notice that the log-likelihood function is depicted as in an ideal situation where estimation is not interfered with by multiple modes or other such problems. The WT focuses on the squared difference between s_1 and s_0 while adjusting for possible difference in curvature between curves a and b. The two likelihood functions are generated by two different sets of data, and the difference in the curvature can be measured by the extent of negativeness of the second derivative of the log-likelihood function. The Wald statistic in a sense "standardizes" the difference $s_1 - s_0$ with a measure of the curvature of the log-likelihood function. In contrast, the LMT measures the square of the score s_0. Similarly, the two curves labeled a and c may have identical slopes but different curvatures. Therefore, we must adjust $[d(s_0)]^2$ by the inverse of $C(s_0)$ to capture the inverse relation between the curvature and the value of the test statistic. The LMT thus gives us another option for hypothesis testing.

2.3 WHAT TO COMPARE?

Oftentimes data analysts seek to compare parameters across groups by testing their hypothesized equality. As will become clear in later chapters, however, this represents only one, albeit arguably, the most common, type of comparison. In addition, one may be interested in comparing across groups the statistical distribution underlying the data, the basic data structure, and the model structure.

2.3.1 Comparing Distributions

Typically the statistical distribution generating a random variable is assumed to be the same in one group as in another group. This, however, need not be true. For example, the distribution underlying the random variable in one group may be normal while the distribution in another may be lognormal. It is also theoretically possible that the distributions generating the random variable in two different groups are two different members of the exponential family when applying the generalized linear model. Or, as in structural equation modeling, the data for onc group may satisfy the multivariate normal assumption while the assumption must be relaxed in another group.

Sometimes the differences between two or more empirical distributions may be of interest to researchers, thereby needing close examination. In Chapter 4 we will study the relative distribution methods, a nonparametric approach, for capturing the differences in empirical distributions.

2.3.2 Comparing Data Structures

When a researcher finds no difference across groups, it is possible that the identity already lies in the basic data structure. For example, when we conduct multiple group analysis in structure equation modeling, the reason that we find no difference in model parameters between two groups may be simply that the correlation matrices of the groups cannot be distinguished. Conversely, the difference in data structure may suggest, though it cannot prove, a difference in parameters. In analyses involving contingency tables, the patterns of sampling zeros may be so different between two groups that they may have laid the foundation for model differences.

2.3.3 Comparing Model Structures

Another level of comparison is the model. It is possible that the relationship between variables can be best represented as one type in one group but as a different type in another group. In generalized linear models, one may entertain the idea of having two different link functions for two different groups, though the link functions rely on the same distribution. This implies that one link function may be canonical and the other not. In structural equation modeling, the data analyst may specify one set of structural relations for group A and a different set of relations for group B. In models with categorical latent variables, the two-way marginals may be modeled with different types of association (as in the log-linear tradition) for different groups.

2.3.4 Comparing Model Parameters

Even in the most common type of comparison there may be variations. Typically we compare parameters or effects of independent variables across groups. When latent variables are present, parameters in the measurement models can also be tested for equality across groups. Thus, measurement parameters, structural parameters, and intercepts in the measurement equations and means (or probabilities) of the latent variables in a model can all be compared across groups. It is the researcher's decision which set of comparisons to make, or which combination of sets of comparisons to make if the purpose is not to compare all parameters there are in a model. Examples in later chapters will demonstrate a number of these possible ways to compare.

A related topic is the comparison of the level of a response variable between groups. While traditionally it is not treated as comparing parameters, one may view such comparison as contrasting intercepts or residual means between multiple groups, and the comparison of the level or mean value of the response variable across groups closely relates to the methods of standardization and decomposition of rates and more generally of any metric variables.

CHAPTER 3

Comparison in Linear Models

3.1 INTRODUCTION

In this chapter we chiefly study linear models of the form

$$\mathbf{y} = \mathbf{X}\boldsymbol{\beta} + \mathbf{e} \qquad \sim N(\mathbf{0}, \sigma^2 \mathbf{I}), \tag{3.1}$$

where $\mathbf{y}$ and $\mathbf{e}$ are random vectors of outcome and error, both of size $N \times 1$, $\mathbf{X}$ is an $N \times k$ matrix containing fixed variables, $\boldsymbol{\beta}$ is a parameter vector of size $k \times 1$, and $\mathbf{I}$ is the unit matrix. However, we will also consider other related methods for comparison, including regression decomposition. We will defer our consideration of nonparametric models until the next chapter.

The OLS estimator of $\boldsymbol{\beta}$ is given by

$$\hat{\boldsymbol{\beta}} = (\mathbf{X}'\mathbf{X})^{-1}\mathbf{X}'\mathbf{y}. \tag{3.2}$$

Alternatively, $\hat{\boldsymbol{\beta}}$ can also be obtained by ML estimation. For the linear model of (3.1), the ML estimator is the same as in (3.2).

This linear model specifies three general statistical methods: analysis of variance (ANOVA), analysis of covariance (ANCOVA), and multiple linear regression. For ANOVA or ANCOVA, $\mathbf{X}$ consists of discrete variables only (for ANOVA), or discrete variables and observed covariates (for ANCOVA). In either case $\mathbf{X}$ is often referred to as the design matrix. For multiple regression, $\mathbf{X}$ contains independent variables predicting or explaining the outcome. These explanatory variables can be either categorical, metric, or both. A major purpose of ANOVA and ANCOVA is to compare means among groups; that can also be one of the purposes of linear regression.

We consider only fixed effects models of groups in this chapter. Such models have explanatory factors regarded as nonrandom, and $\boldsymbol{\beta}$ is a vector of nonrandom coefficients to be estimated. The random effects and random coefficients models will be considered in Chapter 10. Furthermore, we concentrate on situations in which elements of $\mathbf{y}$ are identically and independently distributed (i.i.d.). We postpone as well the discussion of correlated **y** due to repeated or clustered measurements. We also postpone the discussion of comparing odds ratio to the chapter on generalized linear models. Finally, although we focus on a simple, one-factor example in the chapter, the example can be generalized to multiple and crossed factor models.

3.2 AN EXAMPLE

The data in Table 3.1 give the birth weight in grams, the gestational age in weeks, and the sex of 24 babies born in a certain hospital. The male and female babies have similar mean gestational ages (38.333 for males and 38.750 for females). One may raise at least three questions about these data. One may wonder if the average birth weight differs between the sexes, if the average birth weight differs between the sexes given age, and if the rate of growth is the same for the male and female babies. The questions can be answered by analyzing the data with a linear model.

Assuming a linear growth rate, a specific model of (3.1) is

$$y_i = \beta_0 + \beta_1 x_{1i} + \beta_2 x_{2i} + \beta_3 x_{1i} x_{2i} + e_i, \qquad (3.3)$$

where y_i is the birth weight for the ith baby, x_{1i} is the sex (1 if male, 0 if female), x_{2i} is the gestational age, and e_i is the i.i.d. random error with $N(0, \sigma^2)$. Testing the parameter estimates, which represent the test statistic of comparison $\mathbf{s}_g$ in Chapter 2, will enable us to answer the questions, although we will test versions of the model first with either or both of β_2 and β_3 constrained to be zero.

Table 3.1 Birth Weight and Gestational Age by Sex

Male		Female	
Birth Weight	Age	Birth Weight	Age
2968	40	3317	40
2795	38	2729	36
3163	40	2935	40
2925	35	2754	38
2625	36	3210	42
2847	37	2817	39
3292	41	3126	40
3473	40	2539	37
2628	37	2412	36
3176	38	2991	38
3421	40	2875	39
2975	38	3231	40

Source: Dobson (1990, Table 2.3).

3.3 SOME PRELIMINARY CONSIDERATIONS

The first hypothesis is the null one of equal birth weight:

$$H_0: \mu_1 = \mu_2,$$

where s_g is represented by μ_1 and μ_2 respectively for the male and female group-specific mean birth weights. The sample mean for the males is 3,024, and for the females it is 2,911.333.

We will use some nonparametric methods on these data for comparison in the next chapter; for now let us focus on parametric ones. We make the following assumptions for making inference from ANOVA and the other parametric multiple comparison models considered here:

- The e_i's are normally distributed.
- The e_i's are independently distributed.
- The e_i's have same variance σ_e^2 for the groups.

Many traditional tests of homoscedasticity, such as Bartlett's, Hartley's, and Cochran's, are based on normal theory and are very sensitive to the assumption of normality; they tend to report too many significant results for long-tailed distributions. Tests developed in recent decades are less sensitive, and more applicable for a variety of distributions. For example, Levene (1960) proposed an F-test using the i.i.d. normal scores, $z_{ig} = (y_{ig} - \bar{y}_{.g})^2$. Other transformed scores were also considered: $z_{ig} = |y_{ig} - \hat{y}_{.g}|$, $z_{ig} = \ln|y_{ig} - \bar{y}_{.g}|$, and $z_{ig} = |y_{ig} - \bar{y}_{.g}|^{1/2}$. A significant difference between the transformed scores is regarded as evidence of significant differences in σ_g^2. For the birth weight data, Levene's test gives an F of 0.029 with a probability of 0.865, suggesting equal variances.

3.4 THE LINEAR MODEL

Our first model is a simplified version of (3.2):

$$y_i = \beta_0 + \beta_1 x_{1i} + e_i. \tag{3.4}$$

Here we are not concerned about the effect of gestational age on growth or the differential effect of growth by sex. The only question we seek to answer is whether the birth weights for males are different from those for females. When the variables are centered for each group, β_0 is the mean of birth weight for the females (an estimated μ_1), and β_1 gives the difference between the male and female babies. Therefore, μ_2 is estimated by $\Sigma_g \mu_g$ for $g = 1, 2$. Ordinary least squares (OLS) estimates of β_0 and β_1 are 2,911.333 and 112.667, respectively. But β_1 is not statistically significant ($t = 0.977$, $p = 0.339$).

3.5 COMPARING TWO MEANS

Under the assumption of homoscedasticity (i.e., $\sigma_1 = \sigma_2$), the appropriate test is the t-test for comparing two means, a topic treated in every introductory statistics text. Most statistical software packages distinguish between one-sample and two-sample (or independent sample) comparisons. The one-sample procedure compares the mean under examination against some hypothesized location value, whereas the two-sample procedure compares a location parameter such as the mean from two independent samples. Here, as well as in the sections of Chapter 4 dealing with nonparametric tests, we focus on the two-sample (or independent sample) comparisons. Then

$$t_{[n_1+n_2-2]} = \frac{(\bar{y}_1 - \bar{y}_2) - (\mu_1 - \mu_2)}{\hat{\sigma}_{\bar{\mu}_1 - \bar{\mu}_2}} = \frac{\bar{y}_1 - \bar{y}_2}{\hat{\sigma}\sqrt{\dfrac{1}{n_1} + \dfrac{1}{n_2}}}, \quad (3.5)$$

where the null hypothesis of no difference in the mean is tested,

$$\hat{\sigma} = \sqrt{\frac{(n_1 - 1)\hat{\sigma}_1^2 + (n_2 - 1)\hat{\sigma}_2^2}{n_1 + n_2 - 2}},$$

and $n_1 + n_2 - 2$ is the number of degrees of freedom associate with the test. Using the data in Table 3.1, we present a graphic comparison of the means and distribution characteristics by sex (Figure 3.1); we then obtain a t-statistic of 0.977 with 22 degrees of freedom ($p = 0.339$). Obviously, the result is identical to that from the linear model. Despite the appearance of boys weighing 113 grams more than girls on average, the hypothesis of equal weights cannot be rejected statistically.

A related topic is testing the difference between two proportions. Without much complication a variant of (3.5) for the test is

$$t_{[n_1+n_2-2]} = \frac{p_1 - p_2}{\sigma\sqrt{\dfrac{p_1(1-p_1)}{n_1} + \dfrac{p_2(1-p_2)}{n_2}}}, \quad (3.6)$$

where p_g is the observed proportion of events for the gth group, and $p_g(1 - p_g)$ gives its variance. When comparing proportions or rates, oftentimes the compositional structure of the groups under comparison affects the comparison; this entails standardizing one (and sometimes both) of the groups before making the comparison. We will examine this topic in Chapter 5.

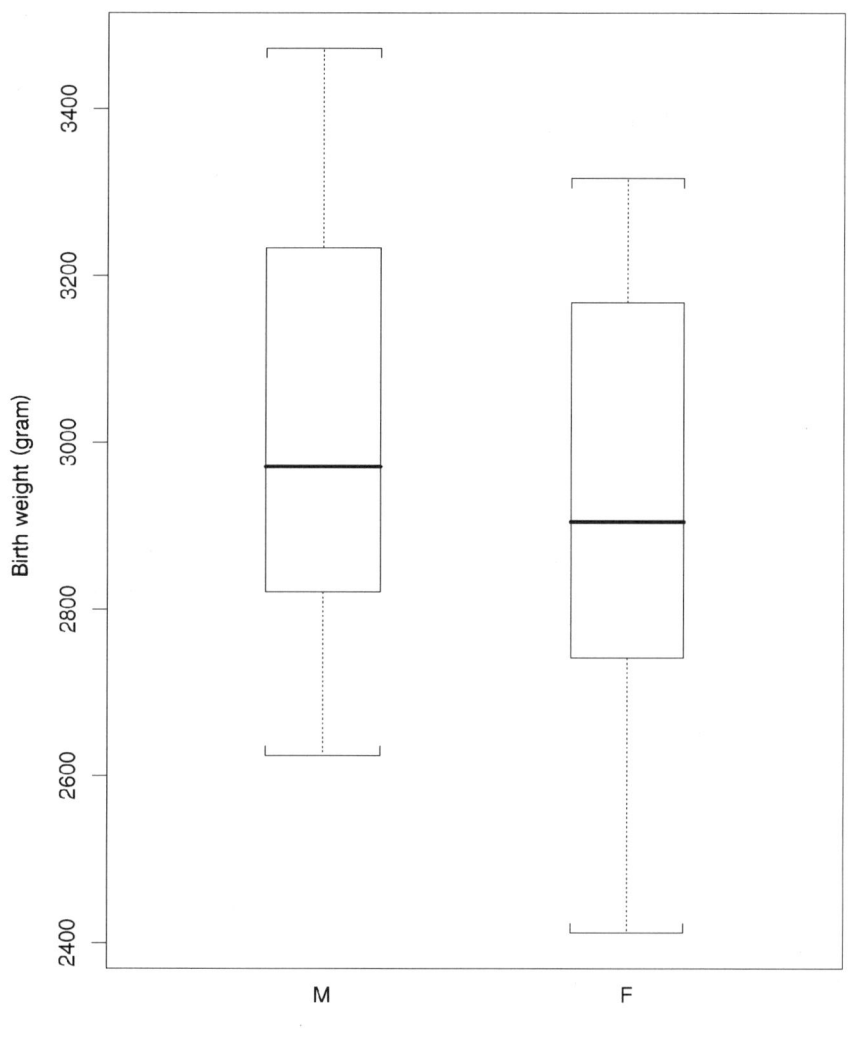

Figure 3.1 A boxplot for the birth weight data.

3.6 ANOVA

The one-factor ANOVA model uses (3.4) as well, and is popular for assessing treatment effects. Here we are interested in whether the response means differ among G groups. Sometimes researchers can randomly assign experimental units to the treatment and control groups corresponding to the levels of a factor. This procedure is termed a completely randomized experimental design. Very often in social science research it is not possible, and quasiexperimental or nonexperimental designs are used instead. Regardless, ANOVA can be applied to assess the difference of an outcome among groups, assuming potential confounding factors have been controlled.

The key concepts in ANOVA are the *between group sum of squares* (SS_B), *within group sum of squares* (SS_W), and *total sum of squares* (SS_T). They are defined as

$$SS_T = \sum_{g=1}^{G} \sum_{i=1}^{n_g} (y_{ig} - \bar{y}_{..})^2, \tag{3.7}$$

$$SS_W = \sum_{g=1}^{G} \sum_{i=1}^{n_g} (y_{ig} - \bar{y}_{g.})^2, \tag{3.8}$$

and

$$SS_B = \sum_{g=1}^{G} n_g (\bar{y}_{i.} - \bar{y}_{..})^2, \tag{3.9}$$

where n_g denotes the total number of cases in the gth group. The relation among the three is given by the identity

$$SS_T = SS_B + SS_W. \tag{3.10}$$

The associated degrees of freedom are $G - 1$ for SS_B, $N - G$ for SS_W, and $N - 1$ for SS_T. By dividing SS_B and SS_W by their respective numbers of degrees of freedom, we arrive at their mean sums of squares, MS_B and MS_W. Using SS_B and SS_W, we may further define three more useful concepts. Unbiased estimators of the variance components σ_e^2 and σ_α^2, are given by

$$\hat{\sigma}_e^2 = SS_W \tag{3.11}$$

and

$$\hat{\sigma}_\alpha^2 = (SS_B - SS_W) n_0, \tag{3.12}$$

where $n_0 = (N^2 - \sum_{g=1}^{G} n_g^2)/N(G - 1)$. For equal group sizes, $n_1 = n_2 = \cdots = n_g$, we have $n_0 = n_g$. From the estimates of σ_e^2 and σ_α^2, the estimate of the total variance σ_y^2 is obtained as

$$\hat{\sigma}_y^2 = \hat{\sigma}_e^2 + \hat{\sigma}_\alpha^2. \tag{3.13}$$

The ANOVA F-test is defined by MS_B/MS_W with the degrees of freedom defined for SS_B and SS_W. For the birth weight data, we obtain an MS_B of 76,162.7 and an MS_W of 79,714.1, giving an $F_{[1, 22]}$ ratio of 0.955 and a probability of 0.339.

3.7 MULTIPLE COMPARISON METHODS

3.7.1 Least Significance Difference Test

This procedure is attributed to Fisher (1935), and has two steps:

1. First, conduct an ANOVA to determine if there are enough significant differences among the groups as suggested by an overall F-test at a chosen α-level of significance.
2. If the F-test is significant, then pairwise comparisons using a t-test are applied to all groups.

When making the comparisons, the statistic of least significant difference (LSD) is used as the criterion:

$$|\bar{y}_g - \bar{y}_h| > t_{[N-G, 1-\alpha/2]} \sqrt{\mathrm{MS}_W \left(\frac{1}{n_g} + \frac{1}{n_h}\right)} \quad \text{for } g \neq h. \quad (3.14)$$

The LSD is given on the right of the inequality. The researcher compares the observed difference of sample means between groups g and h with the value of the LSD. One may conclude that the difference is significant if the sample difference is greater than the LSD.

The data in Table 3.1 give a mean difference of 112.667, but an estimated LSD for the data is

$$t_{[22, 1-0.05/2]} \sqrt{\mathrm{MS}_W \left(\frac{1}{n_1} + \frac{1}{n_2}\right)} = 2.074 \sqrt{\frac{79{,}714.1}{6}} = 239.057.$$

The result shows that the difference, being smaller than the LSD, is not significant.

3.7.2 Tukey's Method

Tukey's method is a simultaneous test for comparing all possible pairwise differences of means of μ_g and μ_h for $g \neq h$ (Tukey 1953). The object is a $100(1 - \alpha)\%$ confidence interval for any pairwise contrast.

The studentized range distribution is at the base of the method. Let r be the range (i.e., maximum−minimum) of some independent observations from a normal distribution with mean μ and variance σ^2, and ν be the number of degrees of freedom for the error ($N - G$ for multiple comparisons). The studentized range is defined as

$$q_{G,\nu} = \frac{r}{\hat{\sigma}}. \quad (3.15)$$

Tukey's method then gives the confidence interval for mean comparions as

$$\bar{y}_g - \bar{y}_h \pm \frac{1}{2}\sum_{g=1}^{G}|c_g|q_{[G,N-G,1-\alpha]}\sqrt{MS_W\frac{1}{n_g}}, \qquad (3.16)$$

where the c_g's are the arbitrary contrasts for the G groups, typically with the constraint of $\sum_{g=1}^{G}c_g = 0$. Since Tukey's method was originally designed for contrasting two means, μ_1 and μ_2, the first term after the $\pm$ sign becomes

$$\frac{1}{2}\sum_{g=1}^{G}|c_g| = \frac{1}{2}(|1| + |-1|) = 1.$$

Although equal group sizes are assumed, Tukey (1953) and Kramer (1956) also developed a modified test where harmonic means of n_g and n_h are inserted for n_g in (3.16). The distribution of q appears in many statistical texts. For example, for the data in Table 3.1 the 95th percentile q is about $q_{[2, 22:0.05]} = 2.94$. This suggests that, if we draw two observations from a normal distribution, the probability is 0.95 that their range is no more than 2.94 times as great as an independent sample standard deviation estimate that has 22 degrees of freedom:

$$\frac{1}{2}\sum_{g=1}^{G}|c_g|q_{[2,22:1-0.05]}\sqrt{MS_W\frac{1}{n_g}} = (1)2.94\sqrt{\frac{79,714.1}{12}} = 239.621.$$

Thus, the 95% confidence interval for the difference in birth weights between the sexes is given by 112.667 ± 239.621.

3.7.3 Scheffé's Method

Scheffé's method also constructs a $100(1 - \alpha)\%$ simultaneous confidence interval for multiple comparisons (Scheffé 1953, 1959). The interval is given by

$$\bar{y}_g - \bar{y}_h \pm f\sqrt{(G-1)MS_W\sum_{g=1}^{G}c_g^2\frac{1}{n_g}}, \qquad (3.17)$$

where $f^2 = F_{[G-1, N-G; 1-\alpha]}$ is the $100(1 - \alpha)$ percentage point of the F-distribution with $G - 1$ and $N - G$ degrees of freedom. For the data at hand, using (3.17) we obtain the 95% confidence limit by

$$\pm\sqrt{f^2(G-1)MS_W\sum_{g=1}^{G}c_g^2\frac{1}{n_g}} = \pm\sqrt{4.30(1)79,714.1(2)\frac{2}{12}} = \pm 338.019.$$

Scheffé's method is more general in that all possible contrasts can be tested for statistical significance and in that confidence intervals can be constructed for the corresponding linear functions of parameters.

As two major methods of multiple comparison, Tukey's and Scheffé's methods have their respective strengths and weaknesses, as summarized below (Sahai and Ageel 2000, p. 77):

1. Scheffé's method can be applied to multiple comparisons involving unequal sample sizes, while the original Tukey's method can only be used with equal sample sizes.
2. Tukey's method is most powerful and gives shorter confidence intervals when comparing *simple pairwise* differences, even though it is applicable for general contrasts.
3. In contrast, Scheffé's method is more powerful and gives shorter confidence intervals for comparisons involving *general* contrasts.
4. If the F-test is significant, then Scheffé's method will detect at least one statistically significant contrast from all possible contrasts.
5. Scheffé's method is more conveniently applicable, because the F-distribution table is more readily available than the studentized range distribution used in Tukey's method.
6. The assumptions of normality and homoscedasticity are more important for Tukey's than for Scheffé's method.

Bonferroni's Method

Bonferroni's method is applicable for a set of pairwise contrasts or linear combinations chosen in advance for an ANOVA with equal or unequal sample sizes. The set cannot be infinite as with Scheffé's method, but may exceed the set of comparisons as in Tukey's procedure. The method is based on the Bonferroni inequality, which states that, if there are M number of null hypotheses and if each one of them is rejected at level α/M, then the overall error rate is at most α. Similarly, if there are M confidence intervals each at the $100(1 - \alpha/M)\%$ level, they should all be true simultaneously with confidence intervals of at least $100(1 - \alpha)\%$. To express the principle mathematically, suppose that each null hypothesis m is tested at level α^* and is rejected with probability R_m; then the overall probability of making a type I error is

$$\alpha = \prod_{m=1}^{M} R_m \leq \sum_{m=1}^{M} R_m = M\alpha^*. \qquad (3.18)$$

The procedure is simple and gives reasonably good results if M is not very large, and tends to be conservative with true confidence intervals greater than $1 - \alpha$. The method should be preferred if there are only a few (say less than G^2) comparisons to be made (Fleiss 1986). For the current data

3.8 ANCOVA

Originally introduced by Fisher (1932), ANCOVA combines ANOVA and regression to assess the statistical significance of mean differences among experimental groups while adjusting the initial observed values of the dependent variable for the linear effect of one or more concomitant variables (covariates). A simple ANCOVA involving one concomitant variable has three steps of computation: we find the sums of squares for the concomitant variable, for the dependent variable, and then for the product of the two. For a mathematical account of the topic, see Scheffé (1959).

One basic assumption of ANCOVA is the homogeneity of group regression coefficients, and that assumption guarantees that the adjustment can be made to any value of the concomitant variables. When the assumption is not made, the effect of the adjustment will be different according to the value of a covariate. The Johnson–Neyman procedure handles the case where the regression coefficients are not homogeneous.

For example, a researcher is interested in studying the effect of a new English textbook on a students' reading ability in one group, compared with a standard text used by another group of students. For good reason we would suspect that socioeconomic status is related to reading. To conduct the adjustment for the only covariate—socioeconomic status, measured by the occupational prestige score of the student's father—we first compute the correlation between the covariate and reading ability, from which we can estimate the amount of variance in reading ability accounted for by socioeconomic status, and the residual variance. This residual variance is used in an ANOVA to estimate the true SS_B after controlling for socioeconomic status.

For the birth weight data, an ANCOVA model gives an MS_B of 157,304 and an MS_W of 31,370, producing an F-statistic of 5.014 ($p = 0.036$). Thus, despite the seeming equal gestational age between the sexes, when the age effect is controlled for, there is significant difference in weight at birth between the sexes.

3.9 MULTIPLE LINEAR REGRESSION

We now extend (3.4) to include gestational age, x_2:

$$y_i = \beta_0 + \beta_1 x_{1i} + \beta_2 x_{2i} + e_i. \tag{3.19}$$

When x_2 is held constant, the difference in y_i becomes statistically significant at $\alpha = 0.05$ with $\hat{\beta}_1 = 163.039$ and standard error 72.808 ($p = 0.036$).

MULTIPLE LINEAR REFRESSION

This result is of course identical to that from the ANCOVA above (notice the identical probability levels). Thus, controlling for age, the estimated sex difference in birth weight is greater than the observed 112.667. The age effect is also highly significant ($\hat{\beta}_2 = 120.894$ with a standard error of 20.463), suggesting a linear growth rate of about 121 grams per week. As a result, the model R^2 increased from 0.042 of (3.4) to 0.640.

So far, we have found no support for the first hypothesis posed at the outset of Section 3.2, but we did find support for the second hypothesis. Now on to the third question: is there differential age effect? This question can be answered by testing a model in the form of (3.3). There is no support for a differential effect of age, however, judging by the t-test for β_3 ($t = -0.441$, $p = 0.664$) or by the F-test for incremental R^2 ($F_{1,20} = 0.168$).

An alternative way to test models with additional interaction terms representing differential effects of certain grouping variables is the Chow test (Chow 1960). This popular method is designed for testing differences between two (or more) regressions, and is based on the assumptions

- $e_{1i} \sim N(0, \sigma^2)$
- $e_{2i} \sim N(0, \sigma^2)$
- $\text{cov}(e_{1i}, e_{2i}) = 0$

That is, the errors in regressions 1 and 2 are distributed normally with zero mean and homoscedastic variance σ^2, and they are independent of each other. With the assumptions satisfied, the Chow test takes the following steps:

1. Run a regression on the combined sample with size N (in the present example, $n_1 + n_2 = N = 24$), and obtain the SS_W and call it S_1. The number of degrees of freedom is $n_1 + n_2 - k$, with k being the number of parameters estimated, including the intercept.
2. Run two regressions on the two individual samples with sizes n_1 and n_2, and obtain their SS_W, or $S_2 + S_3$, with $n_1 + n_2 - 2k$ degrees of freedom.
3. Conduct the $F_{[k, n_1+n_2-2k]}$ test defined by

$$F = \frac{[S_1 - (S_2 + S_3)]/k}{(S_2 + S_3)/(n_1 + n_2 - 2k)}. \quad (3.20)$$

If the resulting F statistic exceeds the critical F, we reject the null hypothesis that the two regressions are equal. Applying the Chow test to our birth weight data, we get

$$F = \frac{[816{,}074.4 - (403{,}698.7 + 248{,}725.9)]/2}{(403{,}698.7 + 248{,}725.9)/(12 + 12 - 4)} = 2.508.$$

The critical value of $F_{2,20}$ at $\alpha = 0.05$ is 3.49. The result is not even significant at the 10% level with a critical value of 2.97). Therefore, we cannot conclude that there is birth weight difference between the sexes controlling for age, and that there is differential growth effect of age due to sex. Figure 3.2 presents the fitted regression lines by sex together with the

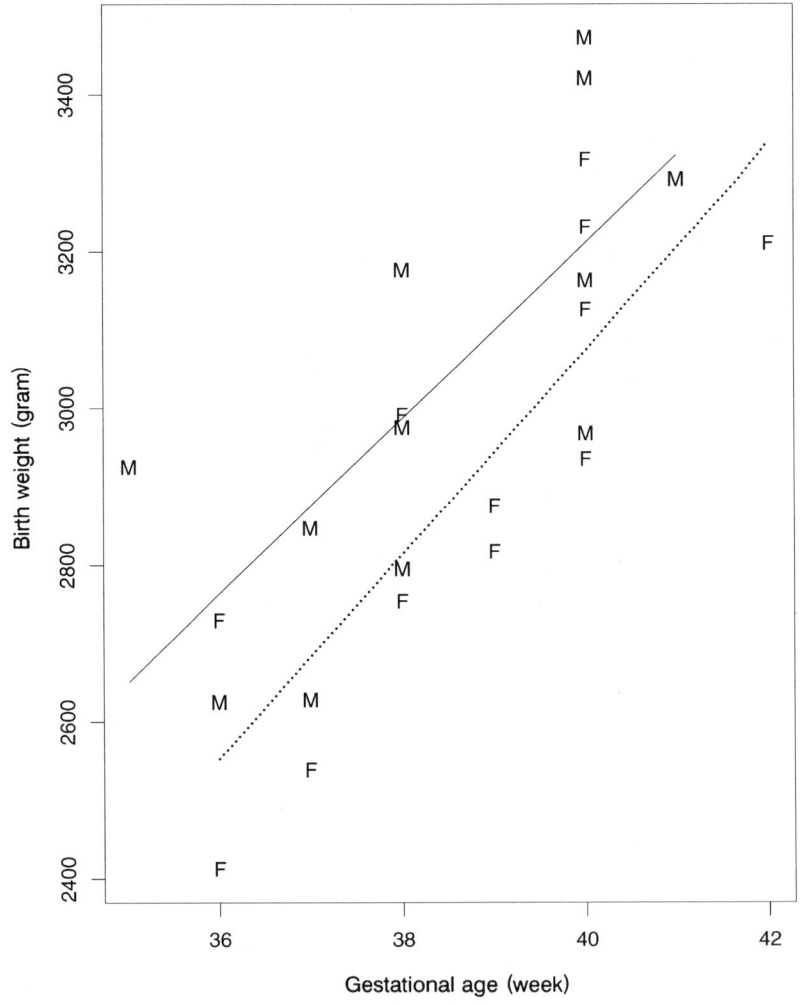

Figure 3.2 A plot of weight by age with regression lines.

observed data points. The visual difference in regression slope between the sexes is minute as well.

Alternatively, we may test the group difference by using an interaction term between sex and gestational age as in (3.3). The idea also applies to the general case with more than two groups and more than one independent variable of interest.

A method for comparison related to multiple linear regression that has fallen out of fashion since its heyday in the 1960s and 1970s is multiple classification analysis (MCA). In MCA there are only categorical explanatory variables, which are effect-coded to deviate from the grand mean. The method can still be quite useful in presenting neatly organized results for comparison. For example, Gershuny (2000) recently analyzed data from 35 surveys in 20 countries, spanning the 1960s to 1990s, to demonstrate the changing time use (in activities such as paid work, domestic work, and leisure) in the world.

A practical issue arising from comparing groups using linear regression analysis is the difference in sample sizes across groups. The Chow test discussed above takes into account the difference in the degrees of freedom of the groups under comparison. Alternatively, some researchers prefer to take a more practical approach to sample size. For example, in Gershuny's (2000) analysis of time use across countries and time periods, he downweighted the number of cases in each country to give a total of 2,000 with approximately 1,000 cases of each sex. This type of weighting can be done in conjunction with other weighting schemes. In his time use research, Gershuny (2000) also weighted the sample of each country at each time point to, say, have comparable socioeconomic composition to that of the U.K. in 1961 (the first time point for the U.K.), rendering the analysis "U.K.-base-year weighted." Readers will find that weighting according to composition is akin to the standardization method introduced in Chapter 5.

3.10 REGRESSION DECOMPOSITION

3.10.1 Rationale

One common use of regression methods is to explain why social, economic, psychological, or medical outcomes differ for different groups. For example, in studying gender differences, we are often interested in determining the components of these differences in social networks, in wage rates, in depression, or in responsiveness to certain medications. Returning to the birth weight example, we may consider two competing explanations or hypotheses:

- Boys tend to have had a longer time to mature in the embryonic stage, and are therefore heavier at birth.
- Girls are genetically predisposed to have smaller stature and thus lighter weight.

These two hypotheses are not mutually inconsistent; they could both be true simultaneously. To assess how much of the difference in birth weight can be explained by the hypotheses, we turn to regression decomposition.

3.10.2 Algebraic Presentation

We rearrange the multiple regression equation of (3.19) to express group as a subscript instead of a variable:

$$y_{gi} = \beta_{0g} + \beta_{1g} x_{1gi} + e_{gi}, \quad (3.21)$$

where g denotes the two sex groups, 1 and 2. We take three algebraic steps to the decomposition of group mean differences. First, we fit a regression through the origin by using group-specific means for y_g and x_{1g}. Now the regression lines go through the point of sample means for the two sex groups (1 = boys and 2 = girls):

$$\bar{y}_1 = \hat{\beta}_{01} + \hat{\beta}_{11} \bar{x}_{11}, \quad (3.22)$$

$$\bar{y}_2 = \hat{\beta}_{02} + \hat{\beta}_{12} \bar{x}_{12}. \quad (3.23)$$

Next, we subtract the mean of group 2 from that of group 1, and obtain

$$\bar{y}_1 - \bar{y}_2 = \left(\hat{\beta}_{01} - \hat{\beta}_{02}\right) + \left(\hat{\beta}_{11} \bar{x}_{11} - \hat{\beta}_{12} \bar{x}_{12}\right).$$

Finally, we add and subtract $\hat{\beta}_{11} \bar{x}_{12}$, and rearrange the terms to obtain

$$\bar{y}_1 - \bar{y}_2 = \left(\hat{\beta}_{01} - \hat{\beta}_{02}\right) + \bar{x}_{12}\left(\hat{\beta}_{11} - \hat{\beta}_{12}\right) + \hat{\beta}_{11}(\bar{x}_{11} - \bar{x}_{12}). \quad (3.24)$$

We now have the difference in the group means expressed as the sum of three terms: $\hat{\beta}_{01} - \hat{\beta}_{02}$, $\bar{x}_{12}(\hat{\beta}_{11} - \hat{\beta}_{12})$, and $\hat{\beta}_{11}(\bar{x}_{11} - \bar{x}_{12})$.

3.10.3 Interpretation

The first term, $\hat{\beta}_{01} - \hat{\beta}_{02}$, gives the difference in the constants, or intercepts. The second, $\bar{x}_{12}(\hat{\beta}_{11} - \hat{\beta}_{12})$, gives the difference in the parameter estimates, weighted by the mean of x_1 of group 2. The third, $\hat{\beta}_{11}(\bar{x}_{11} - \bar{x}_{12})$, gives the difference in the mean value of the explanatory variable, weighted by the group 1 parameter estimate. The estimates of these components are reported in Table 3.2.

What do these components tell us? The difference in birth weight is broken down into three parts. $\hat{\beta}_{11}(\bar{x}_{11} - \bar{x}_{12})$ quantifies the amount of sex difference in birth weight that is due to sex difference in gestational age, expressed in terms of a "standard" effect of gestational age, in this cases the boys'. This terms assesses the maturity hypothesis. $\bar{x}_{12}(\hat{\beta}_{11} - \hat{\beta}_{12})$ expresses

REGRESSION DECOMPOSITION

Table 3.2 Components for Regression Decomposition of the Birth Weight Data

$\bar{x}_{11}$	38.33
$\bar{x}_{12}$	38.75
$\hat{\beta}_{01}$	−1,268.67
$\hat{\beta}_{02}$	−2,141.67
$\hat{\beta}_{11}$	111.98
$\hat{\beta}_{12}$	130.40

the sex difference in birth weight that is due to differences in the effects of gestational age, valued at the mean gestational age for girls. This terms evaluates the genetics hypothesis. Finally, $\hat{\beta}_{01} - \hat{\beta}_{02}$ gives the residual difference in birth weight that is unexplained. If the regression model is believed to have no specification error of omitted variables, then $\bar{x}_{12}(\hat{\beta}_{11} - \hat{\beta}_{12}) + (\hat{\beta}_{01} - \hat{\beta}_{02})$ may summarize the genetic difference. One notices that the decomposition is unique in that if $\hat{\beta}_{11}\bar{x}_{12}$ is replaced by $\hat{\beta}_{12}\bar{x}_{11}$, then group 2 is used as the standard or base group (whose gestational gain is used) for comparison.

The average difference in birth weight of $3,024.00 - 2,911.33 = 112.67$ can now be decomposed into three parts. For the *maturity hypothesis*, $\hat{\beta}_{11}(\bar{x}_{11} - \bar{x}_{12}) = -47.03$ shows the amount of sex difference in mean birth weight that is due to sex difference in gestational age, valued in terms of a "standard" effect of the boys' gestational age. For the *genetics hypothesis*, $\bar{x}_{12}(\hat{\beta}_{11} - \hat{\beta}_{12}) = -713.78$ gives the sex difference in mean birth weight that is due to differences in the effects of gestational age, assessed at the girls' mean gestational age. The *residual effect*, $\hat{\beta}_{01} - \hat{\beta}_{02} = 873.00$, gives the unexplained difference in mean birth weight. The three components sum to 112.19, within rounding error of the mean weight difference of 112.67. In conclusion, genetics contributes over 15 times more than maturity to the weight difference between the sexes, but there is much unexplained effect as well.

3.10.4 Extension to Multiple Regression

We have only examined regression decomposition for a model with one explanatory variable. For a regression model with multiple independent variables, the extension is straightforward, and (3.24) becomes

$$\bar{y}_1 - \bar{y}_2 = \left(\hat{\beta}_{01} - \hat{\beta}_{02}\right) + \sum_{k=1}^{K} \bar{x}_{k2}\left(\hat{\beta}_{k1} - \hat{\beta}_{k2}\right) + \sum_{k=1}^{K} \bar{\beta}_{k1}(\bar{x}_{k1} - \bar{x}_{k2}), \quad (3.25)$$

where subscript k indexes the explanatory variables x, and K is the total number of x's. For regression decomposition involving multiple groups, a natural way is to conduct pairwise comparisons where (3.24) still applies.

3.11 WHICH LINEAR METHOD TO USE?

The many comparative methods we have discussed in this chapter are all based on the linear model. However, they also differ from one another in their statistical foundations. Which method should one use for comparative research?

The very first question to address oneself to is whether the research design is experimental. With an experimental design where randomization avoids any potential confounding effects, a simple t-test for comparing the means should be sufficient. When there are multiple groups form the comparison involving an experimental design, one of the multiple comparison methods in Section 3.5 will be of use. In the same section some (albeit far from complete) comparisons of the multiple comparison methods are offered.

Very often a true experiment cannot be conducted, and quasiexperimental or nonexperimental design characterizes the research. This means there may very well be factors, compositional or not, that may confound the comparison. When all confounding variables are categorical, ANOVA can be used. When there are covariates, ANCOVA can be chosen. Here we only consider one outcome variable for comparison. In situations where multiple (M) outcomes are present, MANOVA and MANCOVA are available.

Multiple regression methods are closely related to ANOVA and ANCOVA, and thus can be used for quasiexperimental and nonexperimental situations where comparison across groups is desired. The choice between regression on the one hand and ANOVA and ANCOVA on the other is largely arbitrary and disciplinary. For example, ANOVA and ANCOVA are more popular in psychology and educational research whereas regression and related methods characterize a lot of sociological and political science research. While multiple regression has been extended to methods that model multiple responses such as seemingly unrelated regressions, for comparisons across multiple groups an understanding for the classical multiple regression should be sufficient for extensions to models for comparison which at the same time take into account correlated multiple outcomes.

CHAPTER 4

Nonparametric Comparison

In this chapter we consider several major nonparametric methods of statistical comparison. These include nonparametric tests such as the Kolmogorov–Smirnov and the Mann–Whitney test, permutation methods, bootstrapping methods, and relative distribution methods. Although not covering every nonparametric method available, these represent some popular conventional nonparametric methods as well as recent developments such as the method of relative distributions.

Albeit all nonparametric in nature, some of these methods can be applied to parametric tests for making statistical comparison. For example, both the permutation and the bootstrap methods can be used to conduct tests such as the t-test or the F-test. The parametric applicability notwithstanding, we include them in this chapter because they do not make the kind of distributional assumptions common in parametric methods in the formation of their theoretical foundation.

4.1 NONPARAMETRIC TESTS

Sometimes parametric tests are inappropriate because their underlying distributional assumptions are questionable. For example, the two-sample t-test assumes that the difference between the samples is normally distributed, whereas a nonparametric test does not make such an assumption. Thus, when the normal distribution assumption does not hold, the nonparametric tests below can be useful. In addition, nonparametric tests do not necessarily require metric data. For example, the Mann–Whitney test can be applied to nonmetric data that are ordered or ranked. Naturally, the outcome variable for the testing in this section is multivalued. For dichotomous outcomes, the McNemar test may be applied.

4.1.1 Kolmogorov–Smirnov Two-Sample Test

This is a test for differences between two distributions, and is based on the unsigned differences between the relative cumulative frequency distributions of the two samples. Critical values can be looked up in a table of the Kolmogorov–Smirnov statistic to determine whether the maximum difference between the two cumulative frequency distributions is significant. The test statistic is given by

$$D = n_g n_h \max \left| \frac{F_g}{n_g} - \frac{F_h}{n_h} \right| \quad \text{for } g \neq h, \tag{4.1}$$

where F_g is the cumulative frequency distribution of group g and F_g/n_g gives the relative cumulative distribution of group g. For the data in Table 3.1, the critical value is 84 (for $\alpha = 0.05$). D is $(12)(12)0.167 = 24$, which is much smaller than the critical value, indicating no significant difference between the two distributions. In the program Stata, D is defined as $\max|F_g/n_g - F_h/n_h|$, and an associate probability is computed (0.996 for example). Some statistical packages such as SPSS also compute a Z-value of the Kolmogorov–Smirnov statistic. For the example, the Kolmogorov–Smirnov Z is 0.408, a rather small value to suggest any significant difference.

4.1.2 Mann–Whitney U-Test

The Mann–Whitney test is equivalent to the Wilcoxon rank sum test. For paired data or one-sample data, the Wilcoxon signed rank test is appropriate. For two-sample data, which is our focus, the Mann–Whitney test (or the Wilcoxon rank sum test) can be performed. The test is based on ranks, and measures differences in location. The test involves the following three steps:

1. Sort the two samples in ascending order.
2. For each observation in one sample (conveniently the smaller sample, say group g), count the number of observations with lower values in the other sample. Let this number be C_{ig} (where i operates within each group g). Ties are treated as $\frac{1}{2}$.
3. The Mann–Whitney statistic U_g is the greater of $\Sigma_i C_{ig}$ and $n_g n_h - \Sigma_i C_{ig}$.

The birth weight data yield a U of 58, with a critical value of 102 ($\alpha = 0.05$), showing no significant difference in ranked birth weight between the two sexes. Critical values for U are given in statistical tables for $n_g \leq 20$. For $n_g > 20$, calculate the quantity t

$$t = \frac{\left(U - \frac{n_g n_h}{2}\right) \pm 0.5}{\sqrt{\frac{n_g n_h (n_g + n_h + 1)}{12}}}, \tag{4.2}$$

where 0.5 is the correction for continuity. When the parenthesized term yields a value > 0, -0.5 is added; otherwise, 0.5 is added. This quantity approximately follows a normal distribution, and uses ∞ for its degrees of freedom. Many statistical packages produce the Mann–Whitney U-test; for smaller samples, you can also use the Web site http://vassun.vassar.edu/~lowry/utest.html for some quick results.

Alternatively, the test statistic can be calculated by

$$U_g = n_g n_h + \frac{n_g(n_g + 1)}{2} - T_g, \qquad (4.3)$$

where T_g is the observed sum rank of group g obtained from a combined ranking of both groups. For the current example, a hand calculation also yields a U-statistic of 58, the same as is generated by SPSS.

4.2 RESAMPLING METHODS

4.2.1 Permutation Methods

Permutation tests can be applied to continuous, ordinal, or categorical data, which may or may not satisfy the normality assumption. They are flexible in that for every parametric or nonparametric test there exists a distribution-free permutation counterpart.

A test ϕ is a decision rule taking values in $[0, 1]$. We reject the null hypothesis if $\phi(x) < p$ with p being a predetermined value, and retain the hypothesis otherwise. A permutation test at α-level consists of a vector of N observations z, a statistic $s(z)$, and an acceptance criterion A, such that for all z, $\phi(z) = 1$ if and only if

$$W(z) = \sum_{\pi \in \Pi} A[s(z), s(\pi, z)] \leq \alpha N!, \qquad (4.4)$$

where Π is the set of all possible rearrangements of the $n_g + n_h$ observations (Good 2000, p. 203).

As a special case of randomization tests, permutation tests have a logic that is easy to understand. A permutation test may take on different forms; for example, it may be combined with a Monte Carlo simulation. It proceeds as follows:

1. Choose a test statistic $s(z)$, and establish its rejection rule.
2. Compute $s(z)$ for the sample, using the original groups of sizes n_g.
3. Combine all groups into one big data set with size N.
4. Conduct permutation:
   (a) Randomly permute the whole sample.
   (b) Divide the resultant sample into groups with sizes n_g.
   (c) Compute $s(z)$ for the new samples.
   (d) Compare the new $s(z)$ with the one based on the original sample.
   (e) Repeat (a) to (d) until ready to make a decision.

5. Reject the null hypothesis if the original $s(z)$ is an extreme value with respect to the rejection rule in the permutation distribution of $s(z)$. Accept the hypothesis otherwise.

If we conduct a permutation test on the birth weight data, we may regard the sex of the babies as unknown, and compare our permuted test results with the original test assuming knowledge of the sex. A decision can then be made with regard to how extreme the original test result is. Alternatively, we may construct a confidence interval based on the distribution of $s(z)$ resulting from all the permutation samples. By the same logic, we can perform t-test and other tests based on the permutation distributions.

To fix the idea, let us look at a very simple example of one outcome variable with cases belonging to two groups. The values of the cases are generated using a random normal distribution with four cases in the first group and two cases in the second. For group 1 the mean is 4 and the variance is 1, and for group 2, they are 2 and 1 respectively. The data, collected in a matrix in S-plus, are displayed as follows:

```
group outcome

[1,]   1 5.140497
[2,]   1 3.416688
[3,]   1 3.581760
[4,]   1 4.505986
[5,]   2 1.235289
[6,]   2 2.364622
```

Using one of the many methods of comparing means discussed in Chapter 3, say linear regression we present the S-plus result of a linear regression model:

```
> summary(lm(outcome~group))

Call: lm(formula=outcome~group) Residuals:

     1        2        3        4        5        6
0.9793  -0.7445  -0.5795   0.3448  -0.5647   0.5647

Coefficients:
              Value Std.   Error t    value     Pr(>|t|)
 (Intercept)    6.5225     0.9885    6.5984    0.0027
       group   -2.3613     0.6990   -3.3782    0.0278

Residual standard error: 0.8071 on 4 degrees of freedom
Multiple R-squared: 0.7405
F-statistic: 11.41 on 1 and 4 degrees of freedom, the
p-value is 0.02783

Correlation of Coefficients:
    (Intercept)
group  -0.9428
```

RESAMPLING METHODS 41

The parameter estimate for the group variable is -2.361, which is the difference in the two group means. (The value can also be obtained by subtracting the mean of group 1 from that of group 2, resulting in -2.361277.) The estimate has a standard error of 0.699, suggesting a rejection of the null hypothesis of no difference at the 3% level.

Here our test statistic $s(z)$ is the difference between the two means, and since the two groups of cases are already combined into one data set, we may proceed with the permutation process. Realistically with larger data sets, we would select random samples of permuted data as the procedure stated above. As we have a rather small data set, we may actually perform a randomization test using the exact permutation distribution that contains all the permutations of the data. Here we create an index variable to indicate the group membership of the cases in all the 15 possible permutations for the cases:

```
>   index
```

	[,1]	[,2]	[,3]	[,4]	[,5]	[,6]
[1,]	1	1	1	1	2	2
[2,]	1	1	1	2	1	2
[3,]	1	1	2	1	1	2
[4,]	1	2	1	1	1	2
[5,]	2	1	1	1	1	2
[6,]	1	1	1	2	2	1
[7,]	1	1	2	1	2	1
[8,]	1	2	1	1	2	1
[9,]	2	1	1	1	2	1
[10,]	1	1	2	2	1	1
[11,]	1	2	1	2	1	1
[12,]	2	1	1	2	1	1
[13,]	1	2	2	1	1	1
[14,]	2	1	2	1	1	1
[15,]	2	2	1	1	1	1

The logic of such permutation is to obtain randomly observed cases for the two groups. Thus the original observations are only one out of 15 possible permutations of the cases.

To compute out permutation statistic of $s(z)$, we create a vector with the length of 15 to contain the $s(z)$'s calculated using the permutation distribution. We then sort the resulting statistics for easier presentation and comparison with the original parametric test statistic:

```
>permstat<- numeric(15)
>for (i in 1:15) permstat[i]<-
+ mean(outcome[index[i,]==2])-mean(outcome[index[i,]==1])
>sort(permstat)
```

    [1] −2.36127708 −1.57222777 −1.44842381 −0.75525455 −0.72522757
        −0.60142361
    [7] −0.27937135   0.09174565   0.18762570   0.56762884   0.88079497
         1.00459892
   [13]   1.35667816   1.48048211   2.17365138

Recall that the original test statistic is -2.361277. By comparing each of the $s(z)$ against the original, we find that only the first $s(z)$ is (in absolute value) greater than or equal to the original test statistic value. This indicates that out of the entire permutation distribution, only the original combination would give a group difference in the mean, thereby yielding $1/15 = 0.0667$. If we choose the conventional 5% level, we will not be able to reject the null hypothesis based on the permutation test.

The software Stata has a user-written module for performing the permutation test. The software can be obtained from http://www.ats.ucla.edu/stat/stata/code/permute.htm. When the sample size is large, samples are taken, and a module like that will be helpful.

4.2.2 Bootstrapping Methods

The bootstrap is another resampling method introduced in recent decades and is computation intensive as well. There are many applications and some textbooks, a good introduction is provided by Efron and Tibshirani (1993). A typical test using the bootstrap proceeds as follows:

1. Choose a test statistic $s(z)$.
2. Compute $s(z)$ for the original sample.
3. Resample with replacement the original sample separately for each original group g, and obtain the bootstrap distribution of $s(z)$ by repeatedly resampling a predetermined number of times.
4. Make a decision of rejecting or accepting the null hypothesis according to the location of the original $s(z)$ with respect to the upper α-percentage point of the bootstrap distribution.

As with the permutation test, the bootstrapping method can also be applied to any parametric statistic of interest.

The key to understanding the bootstrapping method is that it does not assist us in estimating the mean, the difference between means, and other estimates. Instead, it helps us in calculating the standard error and confidence intervals for these estimates, without appealing to asymptotic theory or making arbitrary assumptions about the estimator. Four common types of confidence intervals are the normal, the percentile, the bias corrected, and the accelerated.

Based on the bootstrap estimates obtained using the bootstrap samples, a new standard error is arrived at, and the normal confidence interval is formed accordingly:

$$\text{Prob}(s(z) \in [s(\hat{z}) - t_{1/2\alpha} \times \text{se}(s(\hat{z})), \{s(\hat{z})\} + t_{1/2\alpha} \times \text{se}(s(\hat{z}))]) = 1 - \alpha,$$

where α is the probability of making a type I error. It is clear that once bootstrap samples are drawn and estimation is conducted, normal theory is once again resorted to. To circumvent the parametric trap, one may use the percentile confidence interval, which relies on the empirical distribution of the bootstrap estimates:

$$\text{Prob}(s(z) \in [s(\hat{z})^*_{B \times 1/2\,\alpha}, s(\hat{z})^*_{B \times (1-1/2\,\alpha)}]) = 1 - \alpha,$$

where $s(\hat{z})^*_{B \times 1/2\,\alpha}$ or its counterpart at the upper tail is empirically determined from the bootstrap distribution. Say we draw a 1,000 bootstrap sample, and $s(\hat{z})^*_{B \times 1/2\,\alpha}$ is given by the sample with the 25th lowest valued estimate and $s(\hat{z})^*_{B \times (1-1/2\,\alpha)}$ is given by the sample with the 25th highest valued estimate.

The classic bias estimate in bootstrapping methods is defined as the difference between the mean of the bootstrap estimates and the original sample estimate. More refined bootstrap confidence intervals takes into account bias estimates. For further discussion on the topics of bias-corrected and accelerated confidence intervals, see Efron and Tibshirani (1993).

We analyze below, using bootstrapping methods, the data in the previous section. To simplify estimation, we use a linear regression where the group difference is represented by the parameter for the dummy variable. While most major statistical software packages have some version of bootstrapping in at least one of their procedures (for example, nonlinear regression in SPSS, and Proc Multitest in SAS), Stata has several flexible bootstrapping commands, bstrap, bs, bstat, bsample, to adapt the method to dovetail with any existing estimation methods. We use bs to obtain bootstrap regression estimates:

```
bs "reg outcome group1" "_b[group1]", reps(20)
```

where the value in the last parenthesis gives the number of bootstrap samples requested. Table 4.1 presents the bootstrap estimates and confidence intervals, together with the parametric regression estimates. It appears that after 200 repeated samples the situation does not really improve in terms of smaller values of bias and standard errors. The bootstrap estimation based on 20 samples with replacement yields a confidence interval straddling 0 and a *t*-test not yet significant at the 0.05 level. While all other confidence intervals exclude 0, recall that the true difference from the simulated data should be 2. In this example, the bootstrapped estimations have invariably produced greater standard errors, further confirming the result from the permutation test.

Table 4.1 Bootstrap and Parametric Estimates of the Simulated Data

Sample	$\hat{\beta}$	Bias	se($\hat{\beta}$)	Lower Limit	Upper Limit
Original	−2.361	—	0.699	−4.302	−0.421
20	−2.361	0.568	1.151	−4.770	0.048
50	−2.361	0.286	0.806	−3.981	−0.742
100	−2.361	0.310	0.984	−4.315	−0.408
200	−2.361	0.128	0.844	−4.026	−0.696
500	−2.361	0.200	0.832	−3.996	−0.727
1000	−2.361	0.232	0.887	−4.102	−0.620
2000	−2.361	0.214	0.861	−4.049	−0.673
5000	−2.361	0.205	0.866	−4.059	−0.663

Note: All confidence intervals are 95%. For brevity, only normal confidence intervals are reported.

4.3 RELATIVE DISTRIBUTION METHODS

The recent interest in comparing distributions illustrates another application of nonparametric comparative methods. Oftentimes researchers feel the need for comparing groups in terms of their socioeconomic achievements, educational attainments, and health related behavior or attributes. The focus in comparing the groups may not be restricted to a location parameter such as the mean or the median. Instead, the researcher may be interested in comparing the shape and the overall differences between the distributions. In this section we consider relative distribution methods presented by Handcock and Morris (1998, 1999). Readers interested in the historical roots of the method are referred to these references.

The method is concerned with full distributional comparison based on the relative distribution. The probability density function (PDF) and the cumulative distribution function (CDF) of the relative distribution have been studied independently over the years. Along these lines, the CDF records a score such as test result or income in one distribution to the percentile in another distribution, and the PDF is a rescaled density of the two distributions. Following the notation in Handcock and Morris (1998, 1999), let us denote the CDF of an outcome variable for a reference group by $F_0(y)$. When we compare the outcome of a group at two times, the reference group is the CDF of the original time. When we compare the outcome of two different groups, it is up to the researcher which group is used as the reference. Let us further denote the CDF for the comparison group by $F(y)$. Thus, the comparison can be the group measured at a later time, or another group when we compare groups measured at the same time.

Let Y_0 and Y be random variables sampled from F_0 and F, respectively. The CDFs of F_0 and F are absolutely continuous with common support. The grade transformation of Y to Y_0 is defined as the random variable

$$R = F_0(Y). \tag{4.5}$$

R is real with range $[0, 1]$ and is obtained by transforming Y by the function F_0. The realization of the random variable R is r, and referred to as the relative data. The CDF of R can be represented by

$$G(r) = F\big(F_0^{-1}(r)\big), \quad 0 \leq r \leq 1,$$

where r denotes the proportion of values and $F_0^{-1}(r)$ represents the quantile function of F_0. The PDF of R then is

$$g(r) = \frac{f\big(F_0^{-1}(r)\big)}{f_0\big(F_0^{-1}(r)\big)}, \quad 0 \leq r \leq 1. \tag{4.6}$$

The CDF of the relative distribution is a 45° line when the two distributions are identical. Consequently, the PDF of the relative distribution is uniform.

Relative distribution methods offer straightforward interpretation. One can interpret the relative data as the percentage rank that the original comparison value would have in the reference group, and the relative PDF $g(r)$ as a density ratio of the fraction of cases in the comparison group to the fraction in the reference group at a given level of the outcome $Y(F_0^{-1}(r))$. An example on income to follow will illustrate some features of interpretation when making distribution comparisons.

For the example we examine two European countries—France and Hungary. The data for France are from the 1990 French Household Panel collected among the noncollective households in Lorraine; the data for Hungary are based on the 1992 Hungarian Household Panel with noninstitutional households in the nation as the target population. Removing the minority of missing and zero values on income, the comparative sample yields 2,057 cases for France and 2,030 cases for Hungary.

We first analyze the data by Lorenz (1905) curves and associated Gini coefficients, the common method for making comparisons of income inequality. Lorenz curves graph CDFs and answer the question, "What is the cumulative fraction of Y that is held by the p of the population with lowest values of the variable such as income?" In other words, Lorenz curves contrast cumulative income share against cumulative population share. When absolute equality exists, the Lorenz curve forms a 45° line connecting $(0, 0)$ and $(1, 1)$. Any departure from equality is represented by a curve below this diagonal line. The Gini coefficient ranges from 0 to 1, with 0 representing perfect equality and 1 absolute inequality. Its value corresponds to twice the area between the Lorenz curve and the diagonal. Figure 4.1 presents the Lorenz curves for France and Hungary. The Lorenz curve for France is

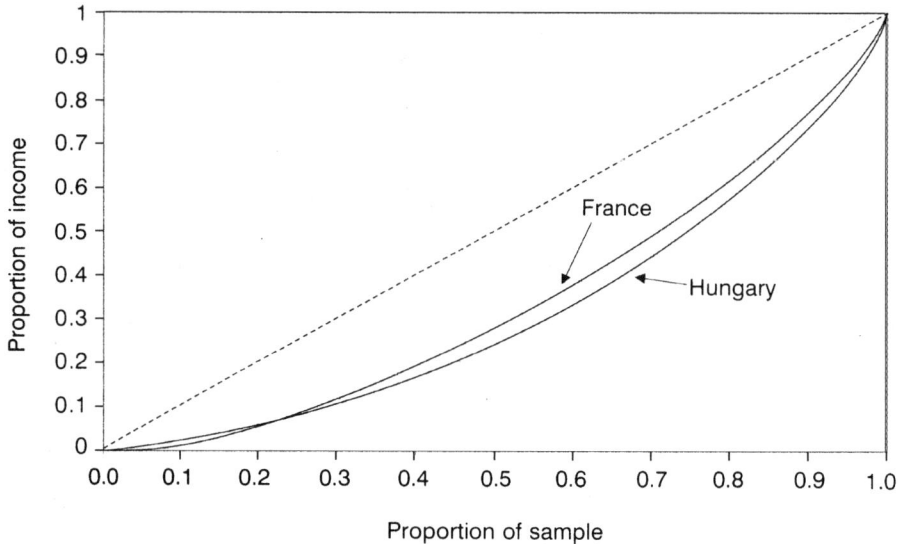

Figure 4.1 Lorenz curves for income distributions of France (1990) and Hungary (1992).

slightly closer to the 45° diagonal, implying less inequality. Similarly, the Gini coefficient for the French data is computed to be 0.322, and that for the Hungarian data, 0.369. While there are many ways to graph Lorenz curves and compute Gini coefficients, a simple way is to create a cumulative Y-variable and a rank-transformed variable of Y on sorted data. Dividing these by the largest cumulative Y-value and the total number of cases, respectively, will result in the cumulative fraction of Y [or $G_L(r)$] and the percentage p of the population variables at each income level. The Gini coefficient can be calculated as

$$\text{Gini}(F) = \frac{N/2 - \sum_{1}^{N} G_L(r)}{N/2},$$

where N represents the total sample size for each population or group. Here we omit the subscript g, as it is understood.

The Lorenz curves in Figure 4.1 support what is suggested by the comparison of the two Gini coefficients: Hungary had a greater amount of inequality than in France in the early 1990s. However, the Lorenz curves also show that the patterns of inequality were different between the two nations. While Hungary had more inequality, it was more pronounced in the middle and higher income levels but less so in the lower income levels than in France.

The Gini coefficient quantifies inequality; the Lorenz curve makes the comparison of patterns of income distribution easier. We will demonstrate relative distribution methods that can give another, more descriptive, and intuitively more appealing perspective on comparing distributions.

Indeed, as Handcock and Morris (1998) pointed out, Lorenz curves are based on relative distributions. The Lorenz PDF is the rescaled density ratio of the currency (*dollar* or *pound*, *franc*, *euro*, etc.), distribution to the income distribution. The currency distribution is the fraction of the total amount of currency in the economy allocated to each income level. Therefore, the Lorenz curve can be viewed as a special form of relative distributions. R_L is the Lorenz transformation of the dollar distribution by the income CDF, and $G_L(r)$, or the Lorenz curve, is the corresponding relative CDF.

As pointed out by Handcock and Morris (1998), there exist three major differences between Lorenz curves and relative distribution methods: the number of groups or populations involved, the unit of measurement, and the level of scale invariance. To begin with the last difference, Lorenz curves are multiplicatively scale invariant in that two distributions will have identical Lorenz curves if they differ only by a multiplicative constant. In contrast, relative distributions are invariant to all *monotonic* transformations of the original measurement. Furthermore, the Lorenz curve graphs cumulative proportions of Y against cumulative proportions of population, and inequality is measured by the departure of the Lorenz curve from the 45° diagonal line, while the relative distribution maps population quantiles to population quantiles for each level of Y. Finally, the Lorenz curve is derived from two attributes of a single group or population, thereby needing two curves to compare two groups, whereas the relative distribution compares one distribution directly with another, using a single curve to reflect the difference between the distributions.

The income data are analyzed using the relative distribution method implemented in the S-plus software, reldist.s, written by Handcock and Morris and available from their web site: http://www.stat.washington.edu/handcock/RelDist/Software/. The software is also available as a SAS macro.

Because of the fluctuation in the currency exchange rate, because of the unavailability of any standard conversion for these income distributions, and more importantly because of our interest in comparing relative distributions and inequality, the Hungarian income variable is multiplied by a constant factor so that the two medians have the same absolute value.

Figures 4.2 and 4.3 show the relative distribution curves comparing the 1990 France and 1992 Hungarian income distributions. Figure 4.2 represents the relative CDF of the income distribution, Hungarian in 1992 to French in 1990. The differences in income between the two countries are described by the horizontal and vertical gridlines. As we set the two medians to be equal,

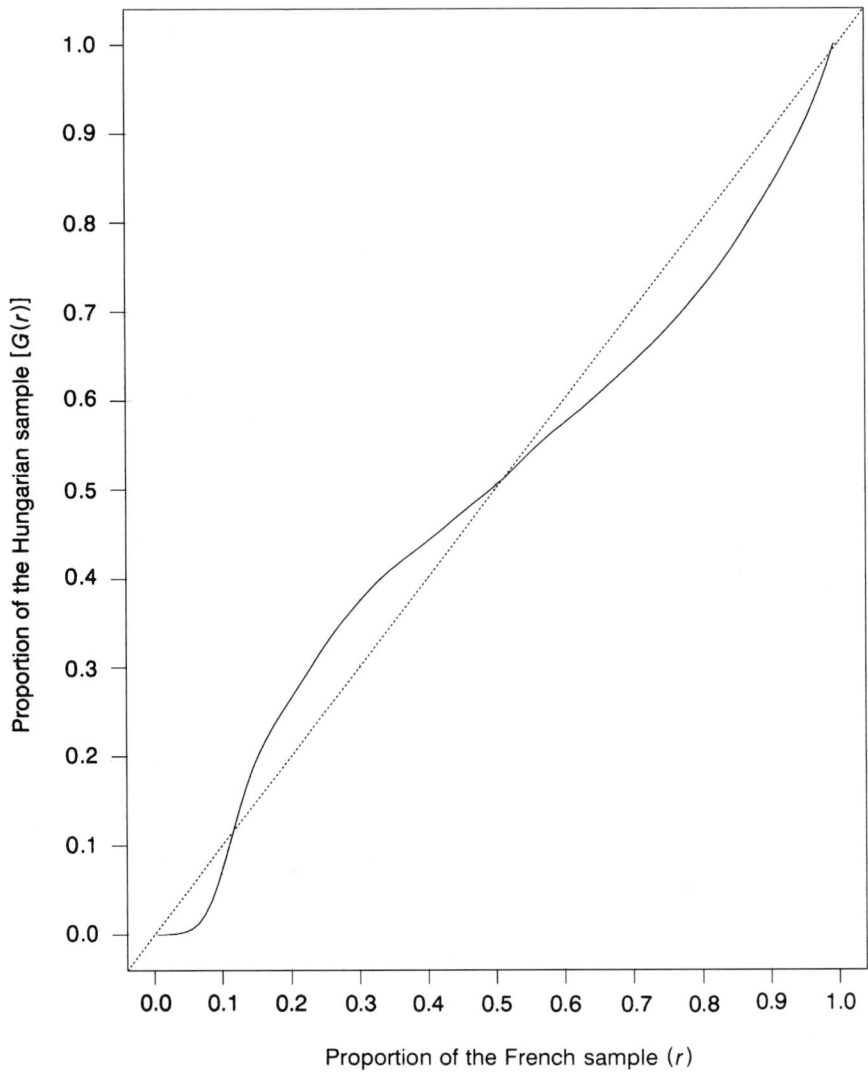

Figure 4.2 Comparison of the relative CDF of income.

$r = G(r)$ at the median. We can see that about 7% of the Hungarian sample is located in the bottom decile of the French sample, whereas about 12% of the French sample is located in the bottom decile of the Hungarian distribution. At the upper end of the distribution, about 84% of the Hungarian sample is located in the ninth decile or below, while close to 95% of the French sample is in or below the ninth decile of the Hungarian distribution.

RELATIVE DISTRIBUTION METHODS

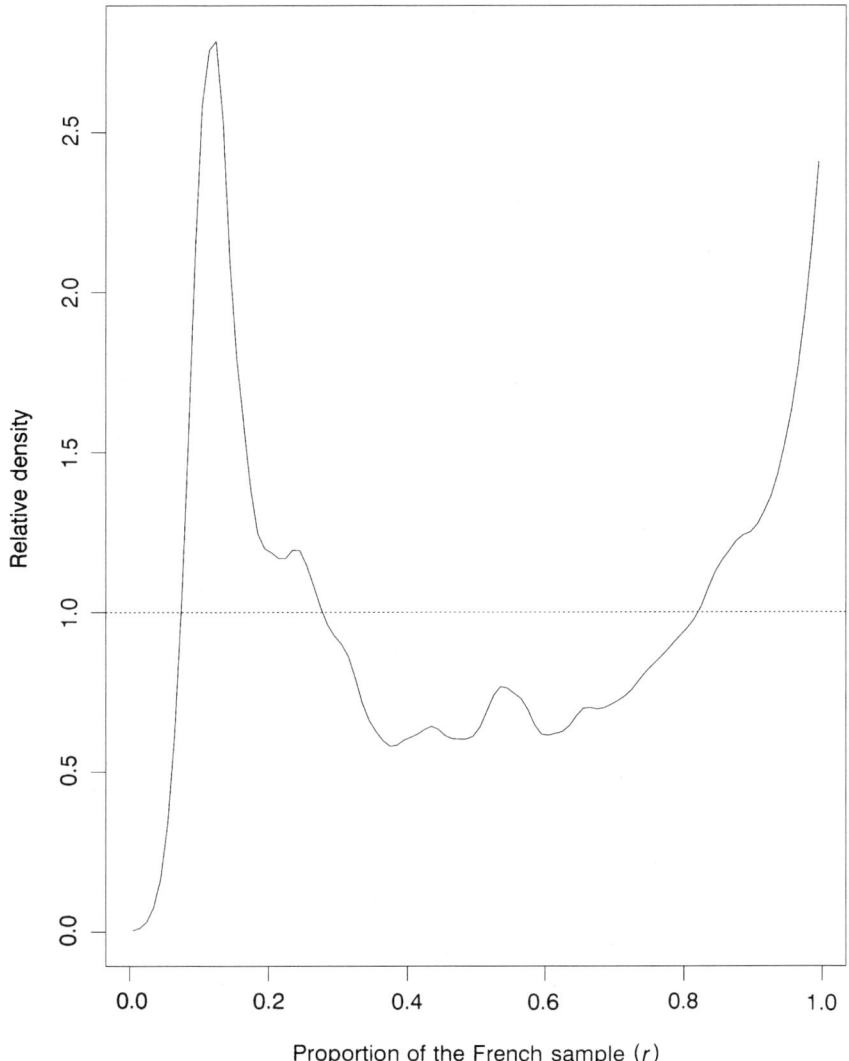

Figure 4.3 Comparison of the relative PDF of income.

The two distributions diverge the most in the third decile and then in the eighth and the ninth deciles. Assuming identical median income levels, there were more Hungarians than French earning an income lower than the median, but more Hungarians than French earning an income higher than the median; hence the tilde-shaped curve. This suggests a greater amount of inequality in Hungary.

While Figure 4.2 describes the relative CDF curve comparing the 1990 French and 1992 Hungarian income distributions, Figure 4.3 shows the relative PDF of the income distribution, 1992 Hungarian to 1990 French. Values above 1, represented by the dotted line, indicate more density in the comparison group, here 1992 Hungarian, than the reference group, here 1990 French. Values below 1 represent less density, with the actual value being the multiplicative factor. It appears that the major differences lie at the bottom and the top of the distributions. Very little income in Hungary fell into the bottom decile defined by the French distribution. However, somewhere between 1.2 and 2.8 times as many Hungarian incomes fall into the French second decile, while fewer Hungarians, as few as half, fall between the 30th and the 80th percentile of the French distribution. But increasingly more Hungarians, as many as 2.5 times, fall above the 82th percentile of the French distribution. Again, the graph shows a greater amount of income inequality in Hongary than in France. This gives a different, perhaps more intuitively appealing, representation of how income inequality compares between the two countries in the early 1990s.

The example only shows one possible way of studying relative distributions. Comparisons using relative distributions can give more information than this. Further analysis can be done in decomposing the relative distribution and in adjusting covariates. For example, the comparison distribution may be a simple shifted version of the reference distribution, and the difference between the two distributions can be summarized by this parsimonious location shift. After adjusting for locational differences, remaining differences can be explained in terms of shape—which commonly consists of spread, skew, and other distributional characteristics. A relative distribution can be decomposed into location and shape differences. Furthermore, the distributional differences can be affected by changes in the distribution of covariates. For example, comparisons of wage distributions can be affected by changes in the distribution of education. The refined relative distribution methods can decompose the relative distribution into components that represent the effects of changes in the marginal distributions of some covariates. For further details on decomposing relative distributions, see Handcock and Morris (1998, 1999).

CHAPTER 5

Comparison of Rates

While a simple comparison of rates can be conducted via a standard t-test identical to the kind described in Chapter 3, sometimes the data contain confounding factors that render a simple comparison invalid if they are not taken into consideration. For example, in demographic as well as biomedical studies mortality and morbidity rates are often compared. It is common knowledge that these rates are not uniformly distributed with age, ethnic groups, and possibly geographic regions. When making comparisons either cross-sectionally or over time, the differences in these composition factors may give rise to differences in the rates even though the pure rates (or composition-specific rates) may not differ much at all. Although a rate may be analyzed as an outcome in a linear model which may further include confounding factors for control, as those treated in the previous chapter, researchers often prefer to treat methods handling rates as a special topic. In this chapter we consider two widely used methods for rate comparison—standardization and decomposition.

5.1 THE DATA

Let us first examine the U.S. mortality data from two time periods—1970 and 1985. It is well known that mortality declined in the U.S. over the 15-year span. However, it is also known that the U.S. population composition has not stayed unchanged either. The data in Table 5.1 give the number of deaths and the size of population classified by age and race. The crude death rates for the U.S. resident population in 1970 and 1985 were 9.422 and 8.739 per 1,000, respectively. On the face of it, the two rates are close to each other, suggesting that mortality in the U.S. did not decline much over the 15 years. However, what may confuse the picture is the changing age structure and racial composition in America, as for several decades the population has been getting older with a larger proportion of nonwhite population. One must adjust mortality rates to take into account this changing population structure.

Table 5.1 U.S. Mortality Data, 1970 and 1985

Year	Age	White Deaths	White Population	Nonwhite Deaths	Nonwhite Population
1970	0	54876	2968	19791	535
	1–4	8624	11484	2924	2162
	5–14	13537	34614	3310	6120
	15–24	35510	30992	9751	4781
	25–34	28286	21983	10907	3096
	35–44	54282	20314	18335	2718
	45–54	138877	20928	30640	2363
	55–64	265133	16897	43240	1767
	65–74	393721	11339	51810	1149
	75–84	456247	5720	33556	448
	85+	232540	1315	14415	117
1985	0	27864	3041	12166	707
	1–4	5351	11577	1988	2692
	5–14	6812	27450	2121	6473
	15–24	30390	32711	7545	6741
	25–34	38463	35480	13389	6547
	35–44	49608	27411	16207	4352
	45–54	92204	19555	24430	3034
	55–64	241237	19795	45243	2540
	65–74	422956	15254	59690	1749
	75–84	513954	8022	54894	804
	85+	389513	2472	29538	236

Note: Population sizes are in thousands.
Source: Liao (1989a, Table 1).

5.2 STANDARDIZATION

A common method to implement adjustments, known as *standardization*, calculates the event rate of a group or population using the statistical information of a standard group or population. This standard can be another population, the same population at an earlier time, or a mathematical average of two (or more) populations. When the statistical information is population composition, the method is known as *direct* standardization; when it is composition-specific rates that are used as the standard for making adjustments, the method is known as *indirect* standardization. For detailed discussion beyond the presentation below see Fleiss (1981).

5.2.1 Direct Standardization

A crude death rate comparison of the two sequences of data in Table 5.1 yields a mortality change of -0.683 ($= 8.739 - 9.422$) per 1,000 population. Did mortality decline this little over the 15 years? To answer the question, we

STANDARDIZATION 53

can calculate new death rates for comparison, assuming a standard population structure. One common choice is a mathematical average of composition proportions. For this example, it makes sense to use the composition proportions of the population at the earlier time as the standard for computing the death rate for the population at the later time. Let us consider the data in Table 5.1 as derived from a contingency table, and consider the observed frequency in cell (i, j, k) of the table. The subscript i refers to a specific category in the population composition; the subscript j refers to a specific group or population for comparison; the subscript k refers to the specific event or state in the outcome variable (life versus death in the current example). The number of people alive of a given age and race group at a given time is simply the size of the population minus the number of deaths. Observed composition-specific rates of event k for group or population j may be defined by

$$r_{ij(k)} = \frac{f_{ijk}}{f_{ij+}}, \qquad (5.1)$$

where the subscript + denotes the marginal total, or summation over the replaced subscript. The crude event rate for the jth group or population is

$$r_{\cdot j(k)} = \frac{f_{+jk}}{f_{+j+}}, \qquad (5.2)$$

where f_{+j+} is the total number of cases in the jth group. For the current example, we are only interested in $k = 1$, and the the sizes of population at risk reported in Table 5.1 are already the sum over $k = 1, 2$. Let us further define the composition proportions as

$$p_{ij} = \frac{f_{ij}}{f_{+j}}, \qquad (5.3)$$

where k is omitted because composition proportions do not concern event categories. A standard proportion p_i^s may be obtained by taking the arithmetic average of p_{i1} and p_{i2} when there are two groups only, as in the current case, or using either p_i^1 or p_i^2 as p_i^s. The directly standardized rate then is

$$r_{\cdot j(k)}^{ds} = \sum_{i=1}^{I} r_{ij(k)} p_{ij}^s, \qquad (5.4)$$

where the summation is over all I number of composition levels i. Applying

(5.4) to the table of f_{ijk} by using p'_i as the standard, we obtained

$$0.009163 \times 0.0.014563 + 0.000462 \times 0.056347 + \cdots$$
$$+ 0.125161 \times 0.000547 = 0.007324.$$

That is, if the population structure in 1985 had remained identical to that in 1970, the death rate would have been 7.324 per 1,000, as opposed to 8.739. Using the earlier population proportions as the standard, the mortality change would be -2.098, as opposed to -0.683 when crude rates are compared. Sometimes it is useful to calculate relative risks. Here the directly standardized relative risk is 0.777 ($= 7.324/9.422$), indicating that, after correcting for age and racial compositional differences, the U.S. population in 1985 had a lower risk of dying than in 1970.

5.2.2 Indirect Standardization

An alternative method is indirect standardization. Suppose that for the later time we have no information on the composition-specific rates. All we have is the population proportions and a crude rate (or the total number of events, deaths in the example). Under such circumstances we can apply indirect standardization, which assumes that the composition structure is the same as observed but that the composition-specific rates are the same as in some standard population.

Using our notation for direct standardization, we first obtain the expected number of deaths assuming the standard composition-specific rates

$$f^{is}_{\cdot j(k)} = \sum_{i=1}^{I} f_{ij+} r^{s}_{ij(k)}, \qquad (5.5)$$

which are taken as the standard for computation. Applying (5.5) to the U.S. mortality data by using the composition-specific rates of 1970 as the standard, we obtain

$$3{,}041{,}000 \times 0.018489 + 11{,}577{,}000 \times 0.000751 + \cdots$$
$$+ 236000 \times 0.123205 = 2{,}648{,}193.$$

Thus, assuming the age and racial structure was the same as observed in 1985 but using the composition-specific rates from 1970 as the standard, the expected number of deaths would be 2,648,193. In mortality or morbidity studies, the results of indirect standardization are sometimes expressed as *standard mortality* (or *morbidity*) *ratios*, which are the ratios of the observed deaths to the expected deaths given the standard composition-specific rates. When the standard rates are taken from one of the groups or populations being compared, the standard mortality ratio becomes a relative risk. For our example the standard mortality ratio is 0.788 ($= 2{,}085{,}563/2{,}648{,}193$), indicating that slightly over 20% more deaths would have been expected in 1985 if their age-race-specific rates had been the same as in 1970.

The indirectly standardized rate is obtained by multiplying the crude death rate by the standard mortality ratio:

$$0.788 \times 8.739 = 6.883.$$

The indirectly adjusted death rate suggests a mortality change of -2.539 ($= 6.883 - 9.422$). Finally, the indirectly standardized relative risk of death comparing the U.S. population in 1985 with that in 1970 is $6.883/8.739 = 0.788$. The inverse relative risk, $1/0.788 = 1.270$, gives the standard mortality ratio, suggesting a 27% higher observed mortality level than what would be expected in 1985 assuming 1970s age-race-specific rates.

One may have noticed that direct and indirect standardization methods give different results. In fact, it is common for different standardization procedures to produce different standardized rates and relative risks as shown above. The whole point is to make adjustments to summary rates for a given group or population, based on some chosen standards, so that comparison is possible. Such standardization facilitates comparison of summary rates of multiple groups but cannot substitute composition-specific rates, which can often be more meaningfully compared.

5.2.3 Model-Based Standardization

The materials in this section assume a basic knowledge of loglinear modeling. Readers unfamiliar with the topic should study it first. The next chapter, on generalized linear models, provides additional materials for understanding it.

The notation we use it, is consistent with that in Sections 5.2.1 and 5.2.2. Consider composition-specific rates of event k for group or population j using model-based, as opposed to observed, frequencies

$$r_{ij(k)} = \frac{F_{ijk}}{F_{ij+}}, \qquad (5.6)$$

where F_{ijk} denotes the expected frequency of the contingency table under some statistical model. The unadjusted event rate for the jth group or population then is

$$r_{\cdot j(k)} = \frac{F_{+jk}}{F_{+j+}}. \qquad (5.7)$$

Here we have a three-way $C \times G \times D$ table with dimensions $I \times J \times K$, where I is the number of composition levels referenced by i, J is the number of groups (G) referenced by j, and K is the number of categories in the dependent variable (D) referenced by k. The composition-specific rates depend on the F_{ijk} in the three-way table cross-classifying C, G, and D. Modeling such a table to assess the association among C, G, and D is the rationale of the purging method (Clogg 1978, Clogg and Eliason 1988, and Clogg, Shockey, and Eliason 1990). A general multiplicative (or loglinear) model for the three-way table is

$$F_{ijk} = \tau \tau_i^C \tau_j^G \tau_k^D \tau_{ij}^{CG} \tau_{ik}^{CD} \tau_{jk}^{GD} \tau_{ijk}^{CGD}. \tag{5.8}$$

The two-factor τ-parameters denote the partial association between pairs of three factors. Of primary interest are the τ^{CG} and τ^{CGD} parameters, which measure the confounding effect resulting from the association between composition and group and the extent to which the group-dependent variable association depends on the levels of composition.

If we disregard the association related to D and instead consider a two-way $C \times G$ table, the model above becomes

$$F_{ij+} = \gamma \gamma_i^C \gamma_j^G \gamma_{ij}^{CG}. \tag{5.9}$$

Here the γ^{CG}-parameters refer to the interaction between composition and group and are called *marginal CG* interactions. The τ^{CG} are called *partial CG* interactions.

The purging method uses the two models above to estimate a set of adjusted or purged rates by using

$$r_{ij(k)}^* = \frac{F_{ijk}^*}{F_{ij+}^*}, \tag{5.10}$$

where the F^*-frequencies are derived by adjusting (dividing) the corresponding F-frequencies with the τ^{CG} and τ^{CGD}, or the γ^{CG}, and so on, whichever is justified by the modeler. The adjusted summary rates are the objective of the purging method, and can be calculated as

$$r_{\cdot j(k)}^* = \frac{F_{+jk}^*}{F_{+j+}^*}, \tag{5.11}$$

where $r_{\cdot j(k)}^*$ represent the adjusted summary rates for each of the j groups.

Four purging methods are defined by the parameters chosen for the adjustment:

Partial *CG* adjustment:

$$F_{ijk}^* = \frac{F_{ijk}}{\tau_{ij}^{CG}}. \tag{5.12}$$

Partial *CG* and *CGD* adjustment:

$$F_{ijk}^* = \frac{F_{ijk}}{\tau_{ij}^{CG} \tau_{ijk}^{CGD}}. \tag{5.13}$$

STANDARDIZATION

Table 5.2 U.S. 1970 Mortality Rates Adjusted Using Various Standardization Methods

Method	Rate 1985	Rate 1970	Difference
Crude	8.739	9.422	−0.683
Direct	7.324	9.422	−2.098
Indirect	6.883	9.422	−2.539
Partial CG	6.939	9.422	−2.483
Partial CG and CGD	6.438	9.422	−2.984
Marginal CG	7.324	9.422	−2.098
Marginal CG and CGD	5.618	9.422	−3.804

Note: The 1970 population is used as the standard for the adjustments for all methods. Computation Software for the purging methods: Eliason's CDAS.

Marginal CG adjustment:

$$F^*_{ijk} = \frac{F_{ijk}}{\gamma^{CG}_{ij}}. \tag{5.14}$$

Marginal CG and three-factor CGD adjustment:

$$F^*_{ijk} = \frac{F_{ijk}}{\gamma^{CG}_{ij} \tau^{CGD}_{ijk}}. \tag{5.15}$$

The purging in these equations creates a set of adjusted frequencies free of association between C and G or among C, G, and D. Now let us apply the methods on the U.S. mortality data. Table 5.2 presents the standardized rates of 1985 with the 1970 population as the standard, using all four purging methods in addition to the direct and indirect standardization results.

The purged rates are computed using the Purge program implemented in Scott Eliason's Categorical Data Analysis System (CDAS), available at http://www.soc.umn.edu/~eliason/CDAS.htm.

As a rule, the marginal CG method gives the closest estimate to the method of direct standardization. For the current example, there is no difference to the third digit after the decimal point. Partial CG gives a result closer to that from the indirect standardization. Partial CG and CGD and marginal CG and three-factor CGD adjustments usually adjust the data table more than the other methods. As previously commented on, there is no *standard* in choosing a standardization method; a different method of standardization and a different standard population will lead to a different set of results.

5.3 DECOMPOSITION

Once rates are standardized, comparison between two rates become possible. However, sometimes we may be interested in finding out exactly how much a rate difference or change is due to compositional difference or change and how much is due to an actual rate difference or change. Decomposition methods go one step further by allocating rate differences into components of rate and various compositional differences.

In her classical research extending standardization to decomposition methods, Kitagawa (1955) demonstrated decomposition when the crude rate difference is confounded by up to three factors, and her two-factor, four-component method has attracted much attention. Many later developments are refinements based on her decomposition methods. A complete review of these methods can be found in Liao (1989a) and Das Gupta (1991). The latter publication also proposed further refinement to Kitagawa's method. While most methods rely on some kind of mathematical adjustment, Liao (1989a) proposed a model-based method. We will examine below this model-based decomposition method in addition to Kitagawa's classical method.

5.3.1 Arithmetic Decomposition

Kitagawa's (1955) two-factor component analysis decomposes the difference between two rates when it is confounded by two factors A and B in the composition C. Factor A has L categories, referenced by l, and factor B has M categories, referenced by m. As before, we have a four-way $A \times B \times G \times D$ table with dimensions $L \times M \times J \times K$, where J is the number of groups (G) referenced by j, and K is the number of categories in the dependent variable (D) referenced by k. Using the notation for observed frequencies in the previous section, we define Kitagawa's four components of rate effect (RE), factor A effect (AE), factor B effect (BE), and the joint effect of factors A and B (ABE) as:

$$\text{RE} = \sum_l \sum_m \frac{\dfrac{f_{lm1+}}{f_{++1+}} + \dfrac{f_{lm2+}}{f_{++2+}}}{2} (r_{lm1(1)} - r_{lm2(1)}), \quad (5.16)$$

$$A\text{E} = \sum_l \sum_m \frac{r_{lm1(1)} + r_{lm2(1)}}{2} \frac{\dfrac{f_{+m1+}}{f_{++1+}} + \dfrac{f_{+m2+}}{f_{++2+}}}{2} \left(\frac{f_{lm1+}}{f_{+m1+}} - \frac{f_{lm2+}}{f_{+m2+}} \right), \quad (5.17)$$

$$B\text{E} = \sum_l \sum_m \frac{r_{lm1(1)} + r_{lm2(1)}}{2} \frac{\dfrac{f_{l+1+}}{f_{++1+}} + \dfrac{f_{l+2+}}{f_{++2+}}}{2} \left(\frac{f_{lm1+}}{f_{l+1+}} - \frac{f_{lm2+}}{f_{l+2+}} \right), \quad (5.18)$$

and

$$ABE = \sum_l \sum_m \frac{r_{lm1(1)} + r_{lm2(1)}}{2}$$
$$\times \frac{\dfrac{f_{lm2+}f_{+m1+}}{f_{+m2+}f_{++1+}} - \dfrac{f_{lm1+}f_{+m2+}}{f_{+m1+}f_{++2+}} + \dfrac{f_{lm2+}f_{l+1+}}{f_{l+2+}f_{++1+}} - \dfrac{f_{lm1+}f_{l+2+}}{f_{l+1+}f_{++2+}}}{2},$$
(5.19)

where in place of the subscripts j and k we simply use 1 or 2 because there are only two groups (populations) for comparison and only one event (death) of concern. Applying these equations to the U.S. mortality data, we obtain an age effect (AE) of 1.566, a race effect (BE) of 0.061, an age and race joint effect (ABE) of -0.081, and a rate effect (RE) of -2.228. These component effects sum to the crude rate difference of -0.683.

5.3.2 Model-Based Decomposition

The four-way table analyzed above can be viewed as a $C \times G \times D$ three-way table because C can always be seen as the cross-classification of multiple factors including A and B. For such a three-way table where the last dimension contains events of concern, the purging method considered in the previous section may be adapted to decompose rate differences. To accomplish this goal, Liao (1989a) extended the purging method to a model-based decomposition method.

Three steps are taken to estimate model-based component effects.

1. A system of simultaneous linear equations is set up as

$$\begin{aligned} \text{RE} + A\text{E} + B\text{E} + AB\text{E} &= \text{CRD}, \\ \text{RE} \phantom{{}+AE} + B\text{E} + AB\text{E} &= \text{RD}(A), \\ \text{RE} + A\text{E} \phantom{{}+BE} + AB\text{E} &= \text{RD}(B), \\ \text{RE} \phantom{{}+AE+BE+ABE} &= \text{RD}(AB). \end{aligned}$$
(5.20)

where RE, AE, BE, and ABE are defined as before. CRD is the crude rate difference. RD(A), RD(B), and RD(AB) stand for the rate differences after the influences of factors A, B, and A and B are purged, respectively. The system of equations is set up so that every term on the left-hand side can be solved by canceling out other terms when one equation is subtracted from another.

2. In the system we need to solve for four unknowns, based on four knowns, of which one is observed and the other three [RD(A), RD(B), and RD(AB)] estimated by purging. Thus, this step involves estimation.

3. Finally, we insert the estimated RD(A), RD(B), and RD(AB) into the equation system to solve for the unknowns.

Steps 1 and 3 are fairly straightforward. Let us focus on step 2 and illustrate the estimation of purged rate differences. To estimate RD(A), RD(B), and RD(AB) using CDAS, input data for the purging method must be arranged accordingly to let C reflect the corresponding factors of A, B, or AB in order to utilize the three-way $C \times G \times D$ setup.

For estimating RD(A) and RD(B), the rate for the factor of interest is adjusted at each level of the left-out variable and pooled proportionately according to the observed frequency distribution of that variable. For example, when the rate for A is being purged of confounding effects, the three-way $A \times G \times D$ table is first evaluated by the purging method at each level of dimension B and then aggregated back to one adjusted summary rate by weighting with the observed proportion of the frequency distribution of B. The rate difference due to factor A is given by

$$\text{RD}(A) = r^{A*}_{\cdot j(k)} - r^{A*}_{\cdot J(k)} \quad \text{for } j \neq J, \tag{5.21}$$

where J is a group whose summary rate is compared with that of the other groups. The simplest situation has only one comparison with $J = 2$. The adjusted rates $r^{A*}_{\cdot j(k)}$ are defined as

$$r^{A*}_{\cdot j(k)} = \sum_m r^{A*}_{m j(k)} \frac{f_{mj+}}{f_{+j+}}, \tag{5.22}$$

where the superscript A denotes that the rate is purged of influence by A. Because the purging is conducted in each of the M three-way $A \times G \times D$ subtables, the adjusted rates must be aggregated back into one adjusted summary rate by applying the weight f_{mj+}/f_{+j+}. This gives rates purged of the influence of A. Similarly, by switching A to B and m to l, we obtain the rate purged of the confounding influence of B, $r^{B*}_{\cdot j(k)}$.

RD(AB), or the rate difference between the groups purged of A and B jointly, can be acquired straightforwardly by inputting $A \times B$ as the composition in C. The rates thus obtained can be written as $r^{AB*}_{\cdot j(k)}$. With these estimated, we are ready to calculate the values of AE, BE, ABE, and RE. A computer program by Albert Chevan implements this method, and is available at: http://staff.uiuc.edu/~f-liao/decomposition/.

We examine below the model-based decomposition method applied to the U.S. mortality data. With standardization it is natural to use a particular group as the reference. With decomposition, however, it is less so, less because we are interested in the components of an *absolute* difference between two rates. In contrast, the focus of standardization is adjusting rates relative to some standard as the reference, hence making the use of a specific group more desirable. Therefore, the decomposition reported in Table 5.3 does not rely on a standard group.

Table 5.3 Component Effects on U.S. Mortality Rate Difference between 1970 and 1985

Method	Crude	Component			
		Age	Race	Joint	Rate
Kitagawa	−0.683	1.566	0.061	−0.081	−2.228
Partial CG	−0.683	2.015	0.040	−0.050	−2.687
Partial CG and CGD	−0.683	2.436	0.017	0.329	−3.464
Marginal CG	−0.683	1.568	0.060	−0.080	−2.231
Marginal CG and CGD	−0.683	1.983	0.037	0.297	−2.999

Note: Components may not sum to the crude rate difference due to rounding errors. Computation Software for the component effects: Chevan's `decompos`.

The component effects derived by using the model-based method should approximate those from more conventional methods of decomposition. For example, the marginal CG purging should give a result comparable to Kitagawa's. Marginal CG purging with no standard group smooths out the interaction between composition and group *only*, just as it approximates the method of direct standardization (Clogg and Eliason 1988). Similarly, Kitagawa's method is designed to smooth out the $C \times G$ interaction by averaging the C proportions in different groups G. The results in Table 5.3 support this observation. The results also show that there is much variation among the methods based on purging. The greatest contrast exists between methods that consider the three-factor interactions and those that do not. This suggests that the CG association is related to the levels of D and is a confounding factor. Notice that the AB joint effect is positive if the three-factor influence is purged. This finding is intuitively appealing, as it implies that the joint age and race work together to cover up the rate difference in addition to the effect of each of the factors alone.

Which method should one use? As Liao (1989a) suggested, if we perceive the problem as a contingency one where dimension C can be collapsed after adjustment, then partial purging is appropriate. If, however, we see C and G as explanatory variables orthogonalizable through purging, then marginal purging can be considered (see also Clogg, Shockey, and Eliason 1990). And if we believe that the CGD interaction should be purged, then one of the methods that model the CGD interaction should be used. For someone who is simply interested in averaging out the differences in the composition, Kitagawa's classic component method will suffice.

The model-based decomposition method can be readily applied to data with more than two composition factors and of more than two groups. For the detailed extension of the method, see Liao (1989a). In general, the model-based approach is more flexible than the conventional method of averaging, especially when there are multiple confounding factors and multiple groups are involved. Chevan's software `decompos` supports rate decomposition of multiple factors and groups.

CHAPTER 6

Comparison in Generalized Linear Models

6.1 INTRODUCTION

This chapter concerns group comparison in statistical models known as generalized linear models (GLMs). A wide variety of models used by applied researchers, including the linear regression, logit, probit, loglinear, and Poisson regression models, are examples of GLMs. McCullagh and Nelder (1989) gives a thorough treatment of the topic, and Dobson (1990) gives a good introduction. The GLM is of the following general form:

$$E(\mathbf{Y}) = \boldsymbol{\mu}, \tag{6.1}$$

$$\boldsymbol{\eta} = g(\boldsymbol{\mu}), \tag{6.2}$$

$$\boldsymbol{\eta} = \mathbf{X}\boldsymbol{\beta}, \tag{6.3}$$

where $\mathbf{Y}$ is an i.i.d. random variable with a probability distribution belonging in the exponential family. The explanatory variables $\mathbf{X}$ and the parameter vector $\boldsymbol{\beta}$ form a linear predictor $\boldsymbol{\eta}$, which is related to $\boldsymbol{\mu}$, the expected vector of $\mathbf{y}$, by a certain link function $g(\cdot)$.

To summarize the GLM differently, we may use a different specification from that in (6.1)–(6.3). Every member

1. The random component: The components of $\mathbf{Y}$ are i.i.d., belonging in the exponential family with distribution-specific parameters describing $E(\mathbf{Y})$ and the variance. For $\mathbf{Y}$ follows a normal distribution with $E(\mathbf{Y}) = \boldsymbol{\mu}$ and constant variance σ^2.

INTRODUCTION

2. The systematic component: The elements $x_1, x_2, \ldots, x_K$ of the vector **X**, combine to make a linear predictor η such that

$$\eta = \sum_{k=1}^{K} \beta_k x_k.$$

The systematic component is identical for all members of the GLM.

3. The link between the random and systematic components, specified by a monotone, differentiable link function $g(\cdot)$: In the case of linear regression models having a normal (or Gaussian) distribution, the link function is identity, so that

$$\eta = \mu.$$

Other link functions take on different forms.

Because component 2 does not change, we will examine the variations in components 1 and 3 in various GLMs in Sections 6.1.1 and 6.1.2 below.

6.1.1 The Exponential Family of Distributions

Let **y** be a realization of **Y**. Each element of **y** has a distribution in the exponential family, taking the form

$$f(y; \theta, \phi) = \exp\left(\frac{y\theta - b(\theta)}{a(\phi)} + c(y, \phi)\right), \qquad (6.4)$$

where $a(\cdot)$, $b(\cdot)$, and $c(\cdot)$ are some functions specific for the member distributions of the exponential family (Barndorff-Nielsen 1978, 1980; McCullagh and Nelder 1989; Nelder and Wedderburn 1972). If ϕ, the dispersion parameter, is known, the distribution is an exponential distribution with canonical parameter θ. If it is unknown, the distribution may or may not be a two-parameter exponential family. The relation between μ and $b(\theta)$ is $\mu = b'(\theta)$. By defining the components on the right-hand side of (6.4), we may define various member distributions of the exponential family.

For example, for the normal distribution, $\theta = \mu$, $\phi = \sigma^2$, $a(\phi) = \phi$, $b(\theta) = \theta^2/2$, and $c(y, \phi) = -\frac{1}{2}[y^2/\sigma^2 + \ln(2\pi\sigma^2)]$, so that

$$f(y; \theta, \phi) = \exp\left[\frac{y\mu - \mu^2/2}{\sigma^2} - \frac{1}{2}\left(\frac{y^2}{\sigma^2} + \ln(2\pi\sigma^2)\right)\right]$$

$$= \frac{1}{\sqrt{2\pi\sigma^2}} \exp\left(-\frac{(y-\mu)^2}{2\sigma^2}\right).$$

This shows how the normal density distribution is related to the exponential family.

For the Poisson distribution, $\theta = \ln \mu$, $\phi = 1$, $a(\phi) = \phi$, $b(\theta) = \exp(\theta)$, and $c(y, \phi) = -\ln y!$, so that

$$f(y; \theta, \phi) = \exp\{(y\ln \mu - \mu) - \ln y!\}$$
$$= \frac{e^{-\mu}\mu^y}{y!} \quad \text{for } y = 0, 1, 2, \cdots, \infty,$$

where μ is the usual Poisson parameter. The Poisson distribution function for the discrete random variable **Y** is shown to be a special case of the exponential family.

Let us see one more example in the binomial distribution. Now $\theta = \ln[\mu/(1-\mu)]$, $\phi = 1$, $a(\phi) = \phi$, $b(\theta) = n\ln(1 + e^\theta)$, and $c(y, \phi) = \ln\binom{n}{y}$, so that

$$f(y; \theta, \phi) = \exp\left\{y\ln\mu - y\ln(1 - \mu) + n\ln(1 - \mu) + \ln\binom{n}{y}\right\}$$
$$= \binom{n}{y}\mu^y(1 - \mu)^{n-y} \quad \text{for } y = 0, 1, 2, \cdots, \infty.$$

where y is the observed number of successes of the random variable **Y** in n independent trials, with the probability or the mean of successes, μ, being identical in all trials. The above demonstrates that the binomial distribution is a member of the exponential family. Other members not discussed here include the exponential, extreme value (Gumbel), gamma, inverse Gaussian, negative binomial, and Pareto distributions.

When ϕ in (6.4) is a known constant, the formula simplifies to that of the natural exponential family

$$f(y; \theta) = a(\theta)b(y)\exp\{yd(\theta)\}. \tag{6.5}$$

We relate (6.5) to (6.4) by $a(\theta) = \exp[-b(\theta)/a(\phi)]$, $b(y) = \exp[c(y, \phi)]$, and $d(\theta) = \theta/a(\phi)$. The two-parameter distributions such as the normal and the gamma with a nuisance parameter ϕ are described by the general formula (6.4), while one-parameter distributions such as the binomial and the Poisson can be more conveniently described by (6.5).

6.1.2 The Link Function

The link function defines the relation between the expected **Y** and η, the linear predictor based on a set of independent variables. Since the link distinguishes the members of the GLM family, and there are many possible

INTRODUCTION

link functions, we focus below on those used in the common types of the GLM after a brief introduction of the concept of a *tolerance distribution*.

Binary outcome models are often used in toxicology to analyze the lethal effect of a certain chemical dosage on a subject. Let x denote the dosage of a chemical, and Y the binary outcome of life or death. Let T stand for the tolerance level of the subject for the dosage; then T can be viewed as an underlying unobserved continuous random variable, and $T - \mathbf{X}\boldsymbol{\beta}$ has a distribution different from that followed by Y. For example, although the outcome data, or the random component of a probit model, follows a binomial distribution, the tolerance distribution for such a model is normal. In that sense, the link function can also be regarded as a link between two different distributions for binary outcome models. With this distinction in mind, we now examine the commonly used link functions.

1. Identity:

$$\eta = \mu.$$

The identity link function defines the classical linear regression model based on the normal distribution.

2. Logarithm:

$$\eta = \ln\mu.$$

The logarithm link is used in Poisson regression models based on the Poisson distribution.

3. Logit:

$$\eta = \ln\left(\frac{\mu}{1-\mu}\right).$$

Applying this link, we specify a logit model that takes on a binary response variable based on the binomial distribution. The model has the logistic distribution as its tolerance distribution.

4. Probit:

$$\eta = \Phi^{-1}(\mu),$$

where Φ^{-1} is the inverse of the standard normal cumulative distribution function. The link is used in a probit model with a dichotomous outcome variable based on the binomial distribution. The tolerance distribution for the model is normal.

5. Multinomial logit:

$$\eta_j = \ln\left(\frac{\mu_j}{1-\mu_j}\right),$$

where j indicates the jth in $1, \ldots, J$ response categories. This link function is a natural extension of the logit link function for a multinomial logit model with a polytomous outcome variable based on the multinomial distribution, which has the binomial distribution as a special case.

6. Complementary log–log:

$$\eta = \ln[-\ln(1-\mu)].$$

This link function is also used in analyzing binomial data, though the tolerance distribution is the extreme value distribution.

For each of the members of the exponential family discussed in Section 6.1.2 there exists a special link function, known as the canonical link, when

$$\theta = \eta,$$

where θ is the canonical parameter defined in (6.4). For example, the canonical links for three of the common distributions are:

1. Normal: $\eta = \mu$;
2. Poisson: $\eta = \ln\mu$;
3. Binomial: $\eta = \ln[\mu/(1-\mu)]$.

For estimating β in the GLM with the canonical link, the sufficient statistics are then $X'y$. This means that $X'y$ summarizes all the available information about, and enough to estimate, β.

6.1.3 Maximum Likelihood Estimation

For brevity let $f(y; \theta)$ denote the joint probability density function

$$f(y_1, y_2, \ldots, y_N; \theta_1, \theta_2, \ldots, \theta_K)$$

based on N observations and K parameters. Because components of Y are independent, the joint probability density function can be expressed as

$$f(y; \theta) = \prod_{i=1}^{N} f(y_i; \theta_1, \ldots, \theta_K). \tag{6.6}$$

Mathematically, the likelihood function $L(\theta; y)$ is the same as the joint probability density function $f(y; \theta)$. However, $f(y; \theta)$ reflects the emphasis on random Y given a parametric model with fixed parameters θ, while $L(\theta; y)$ conveys the idea of how parameters θ are chosen or estimated given the data y.

INTRODUCTION

Let Θ represent the parameter space containing all possible values of the parameter vector $\boldsymbol{\theta}$. The ML estimator of $\boldsymbol{\theta}$ is the estimator that produces $\hat{\boldsymbol{\theta}}$ that maximize the likelihood function, or equivalently, the log-likelihood function $\mathcal{L}(\boldsymbol{\theta}; \mathbf{y}) = \ln L(\boldsymbol{\theta}; \mathbf{y})$. The log-likelihood function is easier to work with. Mathematically,

$$\mathcal{L}(\hat{\boldsymbol{\theta}}; \mathbf{y}) \geq \mathcal{L}(\boldsymbol{\theta}; \mathbf{y}) \qquad \text{for all } \boldsymbol{\theta} \text{ in } \Theta. \tag{6.7}$$

Using (6.4), the log-likelihood function then is

$$\mathcal{L}(\boldsymbol{\theta}, \phi; \mathbf{y}) = \sum_{i=1}^{N} \frac{y_i \boldsymbol{\theta} - b(\boldsymbol{\theta})}{a(\phi)} + \sum_{i=1}^{N} c(y_i, \phi). \tag{6.8}$$

Ultimately, we are interested in obtaining model parameters $\boldsymbol{\theta}$ in the general case, and specifically the regression-type estimator $\hat{\boldsymbol{\beta}}$ is obtained by solving the likelihood equation, which is the first (partial) derivatives of the log-likelihood function set equal to zero

$$\frac{\partial \mathcal{L}(\boldsymbol{\beta}, \phi; y)}{\partial \beta_k} = 0,$$

where $[\partial \mathcal{L} = \partial \mathcal{L}(\boldsymbol{\beta}, \phi; y)]$

$$\frac{\partial \mathcal{L}_i}{\partial \beta_k} = \frac{\partial \mathcal{L}_i}{\partial \theta_i} \frac{\partial \theta_i}{\partial \mu_i} \frac{\partial \mu_i}{\partial \eta_i} \frac{\partial \eta_i}{\partial \beta_k}.$$

Whether the solutions are obtained when the log-likelihood function is at a maxima can be verified by whether the second derivatives of the log-likelihood function are negative,

$$\frac{\partial^2 \mathcal{L}(\boldsymbol{\beta}, \phi; y)}{\partial \beta_k \partial \beta_j} < 0,$$

when evaluated at $\beta_k = \hat{\beta}_k$. An iterative reweighted least squares procedure is used for ML estimation in GLMs. The rate of convergence of $\hat{\boldsymbol{\beta}}$ to $\boldsymbol{\beta}$ depends on the information matrix, which has elements $E[\partial^2 \mathcal{L}(\boldsymbol{\beta}, \phi; y) / \partial \beta_k \partial \beta_j]$ and is equal to

$$I = \mathbf{X}'\mathbf{W}\mathbf{X},$$

where **W** is the diagonal matrix with main diagonal elements

$$w_i = \frac{(\partial \mu_i / \partial \eta_i)^2}{\text{Var}(y_i)}.$$

The information matrix is also used for estimating the variance and covariance of $\boldsymbol{\beta}$.

6.2 COMPARING GENERALIZED LINEAR MODELS

6.2.1 The Null Hypothesis

Consider the following two GLMs for two groups (1 and 2) of observations with a total size of N (which for ease of presentation are sorted into two sets, $i = 1, \ldots, M$ and $i = M + 1, \ldots, N$, respectively):

$$\boldsymbol{\eta}_1 = \mathbf{X}_1 \boldsymbol{\beta}_1, \tag{6.9}$$

$$\boldsymbol{\eta}_2 = \mathbf{X}_2 \boldsymbol{\beta}_2, \tag{6.10}$$

where $\mathbf{X}_1$ and $\mathbf{X}_2$ are $M \times K$ and $(N - M) \times K$ matrices of explanatory variables respectively, $\boldsymbol{\beta}_1$ and $\boldsymbol{\beta}_2$ are $K \times 1$ parameter vectors, and $\boldsymbol{\eta}_1$ and $\boldsymbol{\eta}_2$ are the two linear predictors of dimensions $M \times 1$ and $(N - M) \times 1$ for the two groups. Similar to the null hypothesis of equal means in Chapter 3, the null hypothesis is to test equality between two GLMs

$$H_0: \quad \boldsymbol{\beta}_1 = \boldsymbol{\beta}_2. \tag{6.11}$$

We may generalize the case of two-group comparison to that involving any number of groups G; then the null hypothesis becomes

$$H_0: \quad \boldsymbol{\beta}_1 = \boldsymbol{\beta}_2 = \cdots = \boldsymbol{\beta}_G. \tag{6.12}$$

While the alternative hypothesis for the former is inequality between the two vectors of GLM parameters, that for the latter is any significant departure from equality among the GLM parameter vectors of the multiple groups.

6.2.2 Comparisons Using Likelihood Ratio Tests

For testing the null hypothesis (6.11), we may use a LRT. For the two-group case, the LRT statistic is

$$\text{LRT} = -2(\mathcal{L}_R - \mathcal{L}_U),$$

where

$$\mathcal{L}_R = \sum_{i=1}^{N} \ln L_i(\hat{\boldsymbol{\beta}}, \phi; y_i) = \mathcal{L}(\hat{\boldsymbol{\beta}}) \tag{6.13}$$

for the restricted model, and

$$\mathcal{L}_U = \sum_{i=1}^{M} \ln L_{1i}(\hat{\boldsymbol{\beta}}_1, \phi_1; y_{1i}) + \sum_{i=M+1}^{N} \ln L_{2i}(\hat{\boldsymbol{\beta}}_2, \phi_2; y_{2i}) = \mathcal{L}(\hat{\boldsymbol{\beta}}_1) + \mathcal{L}(\hat{\boldsymbol{\beta}}_2)$$
(6.14)

for the unrestricted model. For testing the hypothesis (6.11), we take the difference of $\mathcal{L}_R$ and $\mathcal{L}_U$:

$$\text{LRT} = -2\mathcal{L}(\hat{\boldsymbol{\beta}}) - \left\{\left[-2\mathcal{L}(\hat{\boldsymbol{\beta}}_1)\right] + \left[-2\mathcal{L}(\hat{\boldsymbol{\beta}}_2)\right]\right\} \sim \chi^2.$$

Thus, the test of equality between sets of parameters in two GLMs involves only subtracting the sum of the −2 log-likelihood functions evaluated at H_1 (the unrestricted model) from the −2 log-likelihood function evaluated at H_0 (the restricted model). That is, the −2 log-likelihood function statistics from the two separate GLMs, one for each of the two groups of observations, are added up and then subtracted from the −2 log-likelihood function statistic from the pooled model. The resultant LRT statistic follows a chi-square distribution asymptotically with degrees of freedom equal to the difference in the number of parameters of the two models. To generalize the LRT above for testing parameter equality among multiple groups as hypothesized by (6.12), we have

$$\text{LRT} = -2\mathcal{L}(\hat{\boldsymbol{\beta}}) - \sum_{g=1}^{G} \left[-2\mathcal{L}(\hat{\boldsymbol{\beta}}_g)\right] \sim \chi^2. \quad (6.15)$$

The test (6.15) is indeed a very straightforward application of the LRT. Note that, as with the ML estimation of parameters in GLMs, this test is also based on the large-sample property.

Alternatively, we may also apply WT. Because the null hypothesis of (6.11), the usual WT formula, under the assumption of fixed X-variables, becomes

$$\text{WT} = (\hat{\boldsymbol{\beta}}_1 - \hat{\boldsymbol{\beta}}_2)'\left[\text{Var}(\hat{\boldsymbol{\beta}}_1) + \text{Var}(\hat{\boldsymbol{\beta}}_2)\right]^{-1}(\hat{\boldsymbol{\beta}}_1 - \hat{\boldsymbol{\beta}}_2) \quad (6.16)$$

where $\text{Var}(\cdot)$ is the estimated variance–covariance matrix of the parameters. Asymptotically, WT is equivalent to LRT, and many software packages readily produce $\text{Var}(\hat{\boldsymbol{\beta}})$.

Another asymptotic equivalent is the LMT. It is also known as the score test, because the gradient vectors used are known as scores. It is given by

$$\text{LMT} = \left(\frac{\partial \mathcal{L}_R}{\partial \hat{\boldsymbol{\beta}}}\right)'\left[I(\hat{\boldsymbol{\beta}})\right]^{-1}\left(\frac{\partial \mathcal{L}_R}{\partial \hat{\boldsymbol{\beta}}}\right), \quad (6.17)$$

where the middle term is the information matrix. The LMT relies on the restricted model only. Because the unrestricted model can also be represented by a pooled model with interactions between $\hat{\boldsymbol{\beta}}$ and the grouping variable and because computers are fast today, a simple way to obtain the score test is to apply the LMT estimation to both the restricted and the unrestricted model and then take the difference of the two score statistics.

The application of (6.15) to comparison using LRT in multiple groups is straightforward, but that of WT or LMT is much less so. In the examples later in this chapter we focus on testing parameter equality between two groups.

6.2.3 The Chow Test as a Special Case

For linear models where the link function is identity, the distribution of the random component is normal, and the error variance is constant, the well-known Chow test can be related to the LRT in Section 4.2.2 as a special case. Nelder and Wedderburn (1972) referred to the LRT statistic as the (scaled) deviance, or D. When the link function is the identity and the distribution is normal, the LRT of (6.14) for testing equality between two groups becomes

$$D = \left[\frac{1}{\sigma^2} \sum_{i=1}^{N} (y_i - \mu_i)^2 - N \ln(2\pi\sigma^2) \right]$$
$$- \left[\frac{1}{\sigma_1^2} \sum_{i=1}^{M} (y_{1i} - \mu_{1i})^2 - M \ln(2\pi\sigma_1^2) \right]$$
$$- \left[\frac{1}{\sigma_2^2} \sum_{i=M+1}^{N} (y_{2i} - \mu_{2i})^2 - (N-M)\ln(2\pi\sigma_2^2) \right]. \quad (6.18)$$

Assuming homoscedasticity or dispersion homogeneity (i.e., $\sigma_1^2 = \sigma_2^2 = \sigma^2$), which excludes any variation of error variance between groups other than through $\boldsymbol{\mu}$, (6.18) simplifies to

$$D = \left[\frac{1}{\sigma^2} \sum_{i=1}^{N} (y_i - \mu_i)^2 \right] - \frac{1}{\sigma^2} \left[\sum_{i=1}^{M} (y_{1i} - \mu_{1i})^2 + \sum_{i=M+1}^{N} (y_{2i} - \mu_{2i})^2 \right]. \quad (6.19)$$

Now let the three SS_W's in (6.19) be designated by D_1, D_2, and D_3 respectively, and regard them as three individual deviance measures. Evidently, $D_1, D_2, D_3 \sim \chi^2$. Since $D = D_1 - (D_2 + D_3)$, dividing this by $D_2 + D_3$ and weighing the numerator and the denominator by their respective

A LOGIT MODEL EXAMPLE

degrees of freedom, we have

$$F = \frac{[D_1 - (D_2 + D_3)]/(2K - K)}{(D_2 + D_3)/(N - 2K)} \tag{6.20}$$

$$= \frac{\left\{\sum_{i=1}^{N}(y_i - \mu_i)^2 - \left[\sum_{i=1}^{M}(y_{1i} - \mu_{1i})^2 + \sum_{i=M+1}^{N}(y_{2i} - \mu_{2i})^2\right]\right\}/K}{\left[\sum_{i=1}^{M}(y_{1i} - \mu_{1i})^2 + \sum_{i=M+1}^{N}(y_{2i} - \mu_{2i})^2\right]/(N - 2K)}. \tag{6.21}$$

(6.20) gives a statistic of an exact test that follows the F-distribution, and is identical to the Chow test given in (3.20) for the classical linear regression under homoscedasticity. To relate (6.21) to (6.15) in terms of D and simplify, we have

$$F = \frac{D/K}{(D_2 + D_3)/(N - 2K)}. \tag{6.22}$$

Therefore, the LRT statistic of (6.15) divided by the sum of -2 times the $\mathscr{L}$'s of the individual GLMs with the identity link, weighted by their respective degrees of freedom, is the same as the Chow test under the assumption of dispersion homogeneity.

6.3 A LOGIT MODEL EXAMPLE

In this section we use the logit model, the most common GLM model with a nonlinear link function, to illustrate testing the null hypothesis of equality between the sets of coefficients in two equations.

6.3.1 The Data

The data in Table 6.1 are a expanded version of the data used by Clogg and Shockey (1988), and contain a binary response recording whether the respondent voted for Reagan or for Carter or another candidate in 1980. In addition to sex, an index of the respondent's political views is the other explanatory variable. The current example, using a sample of 2,910 white men and women who turned out to vote in that year, employs one more variable than the original example and is based on multiple years, as opposed to one year, of the survey. Table 6.1 contains the grouped data for the example.

There appears to be some gender difference in the effect of political views on voting behavior, but nothing is certain until we study the data more formally with some GLM models.

Table 6.1 1980 U.S. Presidential Vote among White Americans

	Male		Female	
Political Views	Reagan	Carter and Others	Reagan	Carter and Others
1	2	21	5	21
2	24	94	41	116
3	58	112	56	122
4	235	216	307	319
5	212	94	161	120
6	197	55	172	69
7	25	17	26	13
Total	753	609	768	780

Source: 1982, 1983, 1984, and 1985 General Social Surveys.
Note: Political views range from 1 = extremely liberal to 7 = extremely conservative.

6.3.2 The Model Comparison

For simplicity and parsimony, a linear logit model,

$$\ln\left[\frac{P(\mathbf{y}=1)}{1-P(\mathbf{y}=1)}\right] = \mathbf{X}\boldsymbol{\beta},$$

which captures the ordinal nature of the categories in political views but assumes a metric measurement, is fitted to the grouped data in Table 6.1. We are interested in the null hypothesis

$$H_0: \boldsymbol{\beta}_m = \boldsymbol{\beta}_f,$$

where $\boldsymbol{\beta}_g$ (for $g = f, m$) contains two parameters for each gender group.

We have estimated four models (see Table 6.2). Model 1 includes all the main effects and the interaction effects between the dummy variable *female* and the other *x*-variable *political views*, thereby representing the conventional dummy-variable approach for testing difference between groups. Models 2, 3, and 4 are identical in the *x*-variable—*political views*—whose effect is being tested for equality between the two gender groups. Model 2 represents H_0—the effect is hypothesized to be equal between women and men—while models 3 and 4 jointly represent H_1 and estimate the effects among women and among men, respectively.

Two general comments are in order. First, the estimates under model 4 are identical to the main effects under model 1, because model 4 has male respondents only, tantamount to setting all variables involving *female* to 0 in model 1. Indeed, adding the estimated effects of *female* and its interactions from model 1 to the estimates in model 4, we would obtain the estimates under model 3. Second, model 2 typically gives a set of "averaged" model 3

A LOGIT MODEL EXAMPLE

Table 6.2 Maximum Likelihood Estimates for Logit Models of the Voting Data

x-Variable	1. One-Model Test	2. Pooled	3. Females	4. Males
Intercept	−2.347	−2.153	−1.963	−2.347
	(0.217)	(0.144)	(0.193)	(0.217)
Female	0.384	—	—	—
	(0.290)	—	—	—
Political views	0.595	0.527	0.463	0.595
	(0.049)	(0.033)	(0.044)	(0.049)
Female × pol. views	−0.132	—	—	—
	(0.066)	—	—	—
Model χ^2	308.783	299.596	23.061	176.366
df for model	3	1	1	1
−2ln L	3,719.344	3,728.531	2,022.830	1,696.514
LRT		9.187		
WT		9.138		
LMT		7.656		

Note: Standard errors are given in parentheses.

and 4 estimates: The estimates from model 2 always fall between those from model 3 and model 4. To graphically compare women and men in their political choices, we present in Figure 6.1 predicted probabilities of voting for Reagan by sex, together with the observed probabilities of the same voting behavior. It appears that model 1 represents the data reasonably well. There is a clear sex difference, indicated by the crossover of the two curves.

The LRT for assessing parameter equality can be calculated easily. Using Equation (6.15), we have 3,728.531 − (2,202.830 + 1,696.514) = 9.187. With two degrees of freedom, which is the number of parameters estimated in models 3 and 4 together minus the number of parameters estimated in model 2, this statistic is significant at the 0.025 level and is almost significant at the 0.01 level. Note that an identical result can be obtained by using the model χ^2's as well. Furthermore, the LRT and the conventional test of joint interaction effects (model 1) will give the same results; researchers may find either method preferable to the other under various circumstances. For example, a researcher who is concerned with the equality of a subset of coefficients only will prefer the dummy-variable approach with interaction terms. However, the approach may become intractable when the number of groups is greater than two and more than a few x-variables are expected to differ across the groups.

Applying Equation (6.16), we get a WT statistic of 9.138, a result almost identical to the likelihood ratio statistic. By using model 1 and constraining the two parameters involving the sex variable to zero (i.e., using the score statistics for models 1 and 2), we obtain a Lagrange multiplier statistic of

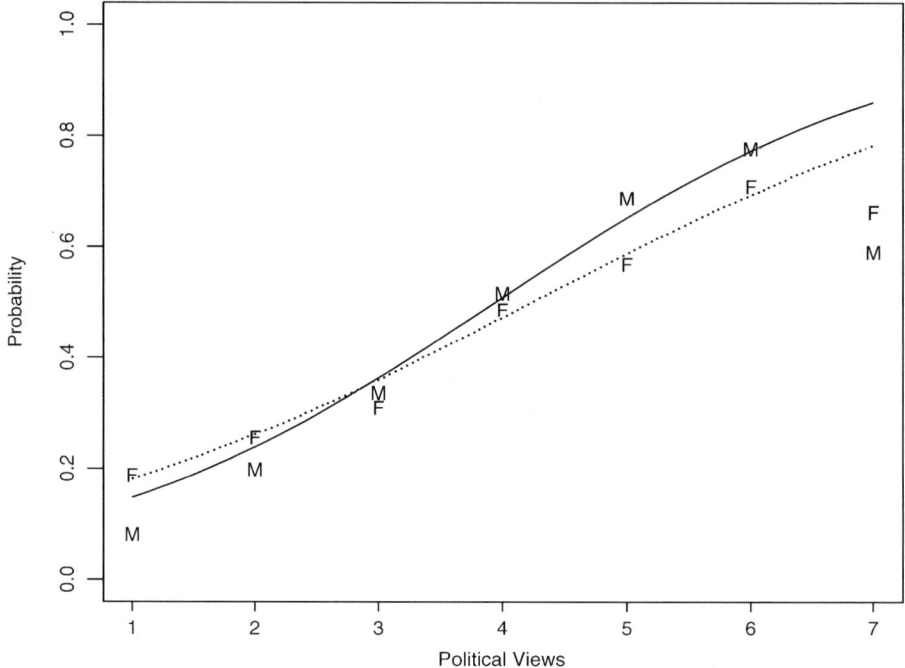

Figure 6.1 Predicted versus observed probabiliyies of voting by sex.

7.656, a result also significant at the 0.025 level. Thus, the three asymptotically equivalent tests give us the same conclusion—a rejection of the null hypothesis of equal parameters between the gender groups.

6.4 A HAZARD RATE MODEL EXAMPLE

6.4.1 The Model

Hazard rate models are used in a type of statistical analysis that goes under different names—event history analysis in the social sciences, failure time analysis in engineering, and survival or duration analysis in the demographic and medical sciences. The reader may refer to any of the many textbooks that describe the fundamentals of the model (Allison 1984; Blossfeld, Hamerle, and Mayer 1988; Blossfeld and Rohwer 1995; Courgeau and Leliévre 1992; Cox and Oakes 1984; Kalbfleisch and Prentice 1980; Lancaster 1990; Lawless 1982; Tuma and Hannan 1984; Vermunt 1997; Yamaguchi 1991). The purpose of this section is only to give a brief introduction to the method in the GLM framework in order to facilitate comparisons.

A HAZARD RATE MODEL EXAMPLE

Several key concepts set hazard rate models apart from other statistical models. First, the dependent variable is defined in terms of the relevant *states* the observations are in. Each observation occupies exactly one of the possible states at a particular point in time. An *event* then is a transition from one state to another. For example, someone who gets married for the first time changes from the state of being never married to the state of first marriage. A *risk set* is composed of all individuals who are at risk of experiencing the event at a particular time, and the period that someone is at risk of a certain event is called a *risk period*. The reason for event history analysis being known also as duration analysis is because it is an analysis of the *duration* of nonoccurrence of an event during the risk period.

Two other important concepts are time varying (or dependent) covariates and censoring. *Time varying covariates* are independent variables that change with time. For example, age and employment status change with time, while ethnic background and sex do not. There are two kinds of *censoring*, left censoring and right censoring. To explain individual differences in the duration of nonoccurrence of a particular event, we need information not only on the covariates determining the process under study, but also on the calendar time of entry into (or beginning of) the risk set (τ_b) and on the calendar time of occurrence of the event (τ_e). The duration of nonoccurrence of an event, T, is simply $\tau_e - \tau_b$. It is possible for some subjects that information on τ_b, τ_e, or both is not available. Left censoring occurs when τ_b is before the observational period, and right censoring occurs when τ_e is after the observational period. Hazard rate models can deal with right-censored data without much complication; left censoring is more complicated (unless τ_b is known).

Let T be a continuous nonnegative random variable representing the duration of nonoccurrence of event, $f(t)$ be the probability density of T, and $F(t)$ be the cumulative distribution function of T. The relation between $f(t)$ and $F(t)$ is

$$f(t) = \lim_{\delta t \to 0} \frac{P(t \leq T < t + \delta t)}{\delta t} = \frac{\partial F(t)}{\partial t}$$

where

$$F(t) = P(T \leq t) = \int_0^t f(s) d(s)$$

and δt is a small time interval. The survival probability or survival function is the probability of event nonoccurrence until time t, and is defined as

$$S(t) = 1 - F(t) = P(T \geq t) = \int_t^\infty f(s) d(s).$$

The *hazard rate* or *hazard function* $h(t)$ measures instantaneous risk, and $h(t)\delta t$ is the approximate conditional probability that an event will occur between t and $t + \delta t$ given that it has not occurred. The hazard rate is defined as

$$h(t) = \lim_{\delta t \to 0} \frac{P(t \leq T < t + \delta t | T \geq t)}{\delta t} = \frac{f(t)}{S(t)}. \quad (6.23)$$

The distribution of, or rather the treatment of the distribution of, the hazard rate defines the types of hazard rate model.

6.4.1.1 Parametric Hazard Models

These models assume a particular functional form for the relationship between T and $h(t)$, and they include many possible functional forms (such as exponential, piecewise exponential, Gompertz, Weibull, log-logistic, log-normal, gamma, and inverse Gaussian), for any of the basic functions $h(t)$, $f(t)$, or $F(t)$.

In GLM terminology, $\boldsymbol{\eta} = \mathbf{X}\boldsymbol{\beta}$, and $\boldsymbol{\eta} = \mathbf{X}(t)\boldsymbol{\beta}$ if the model has some time dependent covariates. Now let $h(t|\mathbf{X})$ be the hazard function at $T = t$ for an observation with covariates $\mathbf{X}$. The exponential model is the simplest parametric hazard rate model. It assumes exponential survival (i.e., time-constant hazard rate),

$$h(t|\mathbf{X}) = \exp(\boldsymbol{\eta}). \quad (6.24)$$

For this model $h(t)$ depends on $\mathbf{X}$ only. In contrast, the Gompertz model extends the exponential model by allowing a monotonic time dependence of $h(t)$,

$$h(t|\mathbf{X}) = \exp(\boldsymbol{\eta} + \alpha t), \quad (6.25)$$

in which $\ln h(t)$ is assumed to be a linear function of T. A further extension is the Weibull model, which assumes $h(t)$ as a log function of T,

$$h(t|\mathbf{X}) = \exp(\boldsymbol{\eta})\alpha t^{\alpha - 1} \quad (6.26)$$

for $\alpha > 0$.

All the parametric models described so far are *proportional* and *loglinear*. They are proportional in that the effect of $\mathbf{X}$ on $h(t)$ is multiplicative, without interaction with time; they are loglinear in that the log of $h(t)$ is a linear function of covariate and time effects. Some parametric hazard models are neither proportional nor loglinear. The log-logistic model is one example:

$$h(t|\mathbf{X}) = \frac{\alpha[\exp(\boldsymbol{\eta})]^{\alpha} t^{\alpha - 1}}{1 + [\exp(\boldsymbol{\eta})t]^{\alpha}} \quad (6.27)$$

for $\alpha > 0$. The log-logistic model can be used for nonmonotonic hazard rates.

6.4.1.2 Cox's Semiparametric Hazard Models

The parametric models require the knowledge of the distribution of T. Cox's semiparametric hazard rate model, also known as Cox's proportional-hazards model, is popular in applied research because it does not assume any particular distributional form of T. The model includes an unspecified function of T in the form of an arbitrary baseline hazard function $h_0(t)$:

$$h(t|\mathbf{X}) = h_0(t)\exp(\boldsymbol{\eta}). \qquad (6.28)$$

The model is parametric in the loglinear relation between $\mathbf{X}$ and $h(t)$. The difficulty in estimating this seemingly simple model is the estimation of $\boldsymbol{\beta}$ without specifying $h_0(t)$. Cox (1972, 1975) proposed partial likelihood estimation that effectively solved the problem.

6.4.1.3 Discrete-Time Models

Sometimes the event time is measured inexactly with only a finite number of time intervals. In such a case we use discrete-time models, which regress the conditional probability of an event during the rth time interval, given that the event has not occurred prior to the interval. We denote that probability by $h(t_r)$. Notice that $h(t_r)$ is only an approximation of $h(t)$.

There are several discrete-time models, the most popular of which applies the logit link function, resulting in

$$h(t|\mathbf{X}) = \frac{\exp(\alpha_r + \boldsymbol{\eta})t^{\alpha-1}}{1 + \exp(\alpha_r + \boldsymbol{\eta})}, \qquad (6.29)$$

which can be reexpressed in terms of the logit:

$$\ln\left[\frac{h(t_r|\mathbf{X})}{1 - h(t_r|\mathbf{X})}\right] = \alpha_r + \mathbf{X}\boldsymbol{\beta},$$

where α_r is a dichotomous variable of the time interval. With some data preparation, the usual logit model can be used on event history data.

6.4.2 The Data

Our data for a hazard rate model come from Krall, Uthoff, and Harley (1975), who analyzed data from a study on multiple myeloma. In the study, researchers treated with alkylating agents 65 patients, 48 of whom died during the study and 17 survived. In the data given in the chapter appendix (Table 6.4), the variable *time* records the survival time in months from diagnosis, and the variable *status* has two values, 0 and 1, indicating whether the patient was alive or dead, respectively, at the end of the study period. Thus, if the value of *status* is 0, the corresponding value of *time* is right-censored.

Table 6.4 contains only two of the nine x-variables—log BUN (blood urea nitrogen, indicator of kidney function) at diagnosis (normal $<$ log 19), and hemoglobin at diagnosis (normal range 14–18 for men and 12–16 for women) —chosen by a forward likelihood ratio selection procedure. These two are the only significant factors related to survival.

6.4.3 The Model Comparison

Comparisons involving hazard rate models follow the same rules as for GLMs as far as discrete-time and parametric models are concerned, because they are all straightforward applications of the GLM. Cox's proportional-hazards model, however, consists of an additional term, $h_0(t)$, which may be subjected to testing of its own hypothesis of equality across groups. For this reason, we present an analysis of applying Cox's regression to the data.

In additional to the hypothesis

$$H_0: \quad \boldsymbol{\beta}_g = \boldsymbol{\beta}_h \quad \text{for } g \neq h,$$

we may also be interested in testing the hypothesis

$$H_0: \quad h_{0g}(t) = h_{0h}(t) \quad \text{for } g \neq h.$$

We conduct a stratified analysis, allowing each stratum to have its own baseline hazard function, but restricting the coefficients to be the same across strata. Most statistical software packages implementing Cox's proportional-hazards model can run stratified analysis. To test the equality of $\boldsymbol{\beta}$, we proceed as before with other GLMs by estimating three models, one pooled and two separate ones for the groups, and then performing a LRT.

We present in Table 6.3 the results from four Cox's proportional-hazards models. The first column contains the model for all cases without stratifying the baseline hazard function by the age variable of two categories. The second has a stratified model of all observations. The third and the fourth

Table 6.3 Comparing Cox's Models of the Data in the Appendix

x-Variable	1. All Cases	2. Stratified, All Cases	3. Age ≤ 55	4. Age ≥ 56
Hemoglobin	−0.119	−0.123	−0.138	−0.116
	(0.058)	(0.061)	(0.151)	(0.066)
log BUN	1.674	1.945	3.441	1.170
	(0.612)	(0.634)	(1.148)	(0.787)
$-2 \ln L, h_0(t)$	309.716	249.711	76.093	173.617
$-2 \ln L, \exp(\boldsymbol{\eta})$	297.767	236.208	65.101	168.273

Note: Standard errors are given in parentheses.

columns are for the models based on the two age groups, those with ages up to 55 and those with ages greater than 55. Even though age was not a significant factor for survival, it does has stratifying effect.

Is the hypothesis $h_{01}(t) = h_{02}(t)$ correct? By taking the difference between the first two of the second line of -2 ln likelihood functions (from the models with the covariates) while the two β-parameters are constrained to be identical for the two groups, we obtain $61.559 = (297.767 - 236.208)$. We may get a similar difference by using the values from the -2 ln likelihood functions for $h_0(t)$ (from the models without the covariates), but the results will not be identical, because the models are different ones when the covariates are absent.

With regard to the second hypothesis, $\beta_1 = \beta_2$, we conclude using (6.15) and the -2 ln likelihood functions for stratified $h_0(t)$ with both identical and different β_g, that the hypothesis is confirmed, because $236.208 - (65.101 + 168.273) = 2.834$ (df = 2). Comparing the parameter estimates across models, we see a big difference in the effect of log BUN between the two age groups. Can this parameter be significantly different between the two groups? Applying (6.16), we obtain a WT statistic of 2.662 (df = 1), an insignificant result. This shows that most of the difference in β is really due to the difference in the effect of log BUN. In sum, the two age groups have significantly different baseline hazard functions but insignificant differences in the covariate effects.

6.A DATA USED IN SECTION 6.4

The data used for the example discussed in Section 6.4 are collected in Table 6.4.

Table 6.4 Survival of Multiple Myeloma Patients after Treatment with an Alkylating Agent

Time	Status	Log BUN	Hemoglobin	Time	Status	Log BUN	Hemoglobin
1.25	1	2.22	9.4	7	1	1.98	9.5
1.25	1	1.94	12	7	1	1.04	5.1
2	1	1.52	9.8	7	1	1.18	11.4
2	1	1.75	11.3	9	1	1.72	8.2
2	1	1.3	5.1	11	1	1.11	14
3	1	1.54	6.7	11	1	1.23	12
5	1	2.24	10.1	11	1	1.3	13.2
5	1	1.68	6.5	11	1	1.57	7.5
6	1	1.36	9	11	1	1.08	9.6
6	1	2.11	10.2	13	1	0.78	5.5
6	1	1.11	9.7	14	1	1.4	14.6
6	1	1.42	10.4	15	1	1.6	10.6

Table 6.4 (*Continued*)

Time	Status	Log BUN	Hemoglobin	Time	Status	Log BUN	Hemoglobin
16	1	1.34	9	88	1	1.18	10.6
16	1	1.32	8.8	89	1	1.32	14
17	1	1.23	10	92	1	1.43	11
17	1	1.59	11.2	4	0	1.95	10.2
18	1	1.45	7.5	4	0	1.92	10
19	1	1.08	14.4	7	0	1.11	12.4
19	1	1.26	7.5	7	0	1.53	10.2
24	1	1.3	14.6	8	0	1.08	9.9
25	1	1	12.4	12	0	1.15	11.6
26	1	1.23	11.2	11	0	1.61	14
32	1	1.32	10.6	12	0	1.4	8.8
35	1	1.11	7	13	0	1.66	4.9
37	1	1.6	11	16	0	1.15	13
41	1	1	10.2	19	0	1.32	13
41	1	1.15	5	19	0	1.32	10.8
51	1	1.57	7.7	28	0	1.23	7.3
52	1	1	10.1	41	0	1.76	12.8
54	1	1.26	9	53	0	1.11	12
58	1	1.2	12.1	57	0	1.26	12.5
66	1	1.45	6.6	77	0	1.08	14
67	1	1.32	12.8				

Source: Krall, Uthoff, and Harley (1975).

CHAPTER 7

Additional Topics of Comparison in Generalized Linear Models

7.1 INTRODUCTION

We consider in this chapter three additional topics of comparison in the GLM—models dealing with dispersion heterogeneity, generalized linear Bayesian models, and models for matched studies. The first two topics are to be viewed as extensions of the usual GLM. The third topic, however, is merely a special application of the GLM. Each of the topics treats the GLM in a new situation of comparison. We discuss the three topics in the following three sections. Matched studies by definition are intended for making comparisons, and in Section 7.2 we examine how such studies can be handled by the GLM. The usual GLM assumes that the dispersion parameter ϕ is constant. However, some data may not satisfy this assumption. In Section 7.3, we present ways for dealing with dispersion heterogeneity. The GLM is based on frequentist statistics. In Section 7.4, we inject some Bayesian thinking into comparison in GLMs.

7.2 GLM FOR MATCHED CASE–CONTROL STUDIES

Case–control designs are widely used among researchers in epidemiology and medicine where a study may be either retrospective or prospective. Recently, however, case–control methods have been advocated for use in sociological research to study rare events (Lacy 1997). Smith (1997) discussed matching with multiple controls to estimate treatment effects in observational studies in the social sciences. Because true experiments are difficult to conduct in many of the social sciences, a matched case–control design may serve as a viable alternative for controlling confounding factors.

First, let us draw the distinction between a retrospective and a prospective study. Two criteria are how the sample is chosen and how the information is collected.

- In a *retrospective* study, we draw a sample according to the present outcome and explore the values of previous explanatory, and possibly outcome, variables. Many case–control studies are retrospective.
- In a *prospective* study, we draw a sample according to the selection criteria for certain explanatory variables, and follow up to see what outcome may occur. Clinical trials and cohort studies are often prospective.

Case–control methods relate to but are distinct from cohort methods in important ways. A cohort study has as its starting point the recording of all subjects, exposed or not. Both those exposed (index subjects) and those not exposed (control subjects) are followed in the same way over time. Depending on whether the data are gathered after or before the event, a cohort study can be retrospective or prospective. In contrast, the starting point of a case–control study is the recording of some subjects exhibiting certain outcome (cases). A comparison group consists of individuals not exhibiting that outcome (controls), and the histories of these subjects are recorded in the same way as those of the cases. Such a design is retrospective, but a case–control study can be prospective as well.

Common to both the cohort and the case–control methods are potential confounding factors. That is why oftentimes case–control studies match on certain confounding factor(s). Analysis of unmatched case–control data can proceed with the GLM for binary data without complication. Thus, here we focus on the statistical analysis of matched case–control data.

Three designs often used are case–control studies of $1:1$, $1:m$, and $n:m$ match. The $1:1$ matched case–control study has one case and one control in each pair; we may regard the concept of *pair* as corresponding to that of the *risk set* in survival analysis. The $1:m$ design has one case and m controls in each set, while the $n:m$ design has n cases and m controls in each set. In a matched case–control study, the status of being a case or a control is treated as an outcome, and we are interested in comparing the cases and the controls with regard to whether the observation is in the treatment (exposure) group or in the group not exposed to the treatment. With a 2×2 design, a conventional way for testing the treatment effect will be that for rates or odds. One such conventional test is the McNemar (1947) test. Since there are often other explanatory variables of interest than just treatment, most often today's researchers use the conditional logistic regression.

In matched case–control studies, it is necessary to introduce a new parameter for every case–control pair or set. These parameters are known as *nuisance parameters*, because they are of little interest to the researcher but complicate the estimation. To eliminate these nuisance parameters from the estimation, the *conditional likelihood*, a modified likelihood function, is used. The conditional logistic regression is based on conditional likelihood estimation.

In a typical matched case–control study, we are interested in testing the hypothesis

$$H_0: \quad \boldsymbol{\beta}_1 = \boldsymbol{\beta}_0$$

for the model

$$\ln\left[\frac{P(y_1 = 1)}{1 - P(y_1 = 1)}\right] = \mathbf{X}_1\boldsymbol{\beta}_1, \quad \ln\left[\frac{P(y_0 = 1)}{1 - P(y_0 = 1)}\right] = \mathbf{X}_0\boldsymbol{\beta}_0 \quad (7.1)$$

where $\boldsymbol{\beta}_1$ and $\boldsymbol{\beta}_0$ are the parameter vectors for $\mathbf{X}_1$ and $\mathbf{X}_0$, respectively, which usually contain in the first column the status of treatment group. The subscript 1 or 0 refers to whether the observation is a case or a control.

Notice that $\mathbf{X}_1$ and $\mathbf{X}_0$ may be of different lengths. That is, although they have an identical number of rows for the 1:1 design, $\mathbf{X}_0$ contains at least an equal but often a greater number of rows than $\mathbf{X}_1$ for the $1:m$ and $n:m$ designs. Assuming independence of responses for different matched sets and for the two groups of responses of the same set, a joint likelihood function can be written out, as for any logistic regression models. In a case–control study that is not matched, these two input matrices as well as the vectors $\mathbf{y}_1$ and $\mathbf{y}_0$ may be vertically concatenated. But that approach will not work here. We demonstrate below how the conditional logistic regression is applied for the $1:1$, $1:m$, and $n:m$ case–control studies.

7.2.1 The 1:1 Matched Study

When there is only one control for each case, we have a simple situation for applying the conditional logistic regression, because the $\mathbf{y}_1$ and $\mathbf{y}_0$ as well as $\mathbf{X}_1$ and $\mathbf{X}_0$ are of the same dimensions respectively. The null hypothesis here is that cases and controls are equally exposed. In other words, there is no treatment effect. When $\mathbf{X}$ has only one factor measuring treatment–exposure status, the *McNemar test* is an option. The observed counts fall into the four cells of a 2×2 contingency table (Table 7.1).

If the null hypothesis is true, then the expected frequencies of the discordant pairs should be equal to the average of the observed frequencies, that is, $F_{12} = F_{21} = (f_{12} + f_{21})/2$. Following this logic and simplifying, a

Table 7.1 Data from 1:1 Matched Pairs

Case	Control	
	Exposed	Not exposed
Exposed	f_{11}	f_{12}
Not exposed	f_{21}	f_{22}

chi-square statistic expressed in observed counts only is

$$\chi^2 = \frac{(|f_{12} - f_{21}| - 1)^2}{f_{12} + f_{21}}. \tag{7.2}$$

This is the McNemar test, which has a df of 1.

More generally, when there are more than one factor in **X**, the regular logistic regression procedure in most statistical software packages may be employed for analyzing such data. The application involves three steps:

1. Take the difference of $\mathbf{X}_1$ and $\mathbf{X}_0$ by forming $\mathbf{X}_d = \mathbf{X}_1 - \mathbf{X}_0$. This effectively transforms individual data into pair data, with each column of $\mathbf{X}_d$ recording the difference of every explanatory variable including the treatment status between cases and controls.
2. Use for the new outcome variable a constant 1. Equivalently, taking the difference of $\mathbf{y}_1$ and $\mathbf{y}_0$, too, produces a vector of 1's with N rows (N being the number of sets or pairs). That is, $\mathbf{y}_1 - \mathbf{y}_0 = \mathbf{1}$.
3. Fit a logistic regression with no intercept. After step 2, the intercept is no longer necessary for estimating the mean of the response.

The procedure is very simple and straightforward to apply. However, certain software such as SPSS (version 10) does not give the flexibility of fitting a logistic regression with a constant outcome variable.

7.2.2 The 1 : m Design

The 1 : m design describes the situation where there is more than one control per case in a matched set. We will postpone the consideration of estimation of the conditional logistic regression for a 1 : m design with a varying m until the next section where the more general case of n:m studies are discussed. Instead, we focus on a special case of the 1 : m design, where the number of controls is constant for all sets and where there is only one covariate, the exposure variable, as discussed by Breslow (1982).

The regular logistic regression can still be applied to the special case. However, the procedure is different from that for the 1 : 1 design. It involves three steps:

1. Arrange the data in terms of observed number of sets (n) by the total number of subjects (both case and controls) being exposed to the treatment (t) in each set. Out of the total number t, the number of sets that are exposed is recorded as the variable e.
2. Compute an offset variable o by $o = \ln[t/(m + 1 - t)]$.

3. Compute a new treatment–exposure variable $x = 1$.
4. Fit a logistic regression model of e events out of n trials with the offset variable o and the constant x but no intercept.

In the Breslow data, there are four controls for each case, and there are 80 sets altogether. In tabular form, the data are

t	0	1	2	3	4	5
e	—	5	19	10	6	0
n	10	20	27	17	6	—

The offset variable $o = \ln[t/(5-t)]$. The first missing value in e results in ten unusable sets, and there is no observation of the combination of five treatments (regardless of case or control) out of five trials. The data can be analyzed by any standard logistic regression procedure, as the programmers of SAS demonstrated in their sample programs.

7.2.3 The $n:m$ Design

When one or more cases are matched with one or more controls in every set, we have the more general $n:m$ matched case–control design, which has both the $1:1$ and the $1:m$ designs as special cases. To see how an analysis of such data works, let us examine some sample data.

7.2.3.1 The Data
The data set from the SAS sample library contains 174 women, 54 of whom had low-birth-weight (LBW) babies (cases), while 120 had normal-birth-weight babies (controls). Several risk factors (covariates) worth examining are weight in pounds at last menstruation (WLM), smoking during pregnancy (SDP), presence of hypertension (HT), and presence of uterine irritability (UI). Age is the confounding factor that serves as the matching variable. All variables except age and WLM are dummy. The case–control variable LBW is coded 1 if case and 0 if control. The observations are ordered by case–control status in ascending order of age. The data are reported in the chapter appendix.

7.2.3.2 The Comparison of Cases and Controls
For the data in Appendix 7.A we are interested in testing

$$H_0: \quad \beta_1 = \beta_0,$$

where each β-vector contains the parameters for WLM, SDP, HT, and UI while controlling for the confounding factor of age, which has 17 levels. To estimate a conditional logistic regression model of the $n:m$ matched data, short of writing our own computer subroutines or macros, we can either use a procedure written for such model (e.g., Stata) or use an existing computer procedure for Cox's proportional-hazards model (e.g., S-Plus, SAS, or SPSS) or for generalized linear models (e.g., Glim). The partial likelihood estimation for Cox's regression essentially takes care of the computation of conditional likelihood in conditional logistic regression models. We demonstrate below how to estimate such a conditional logistic regression model with a Cox's regression procedure.

To use a Cox's regression procedure for estimating conditional logistic regression models, there are four steps:

1. Recode the case–control variable into a failure time (event time) variable with the value 1 for cases and 2 for controls. In the current example, this can be achieved by time = 2 − LBW.
2. Either create a new variable, *censor*, that has a value of 0 for cases and 1 for controls, or use the event time variable. The most important thing is to declare the correct censoring value (1 for censor or 2 for time in the current example).
3. Use the matching variable(s) as the stratifying variable(s). You may bring them directly into the model specification as stratifying variable(s). Some software packages only allows the inclusion of one stratifying variable. In that case, you must code the matching variables into a new risk-set variable. This can be done by cross-classifying the original matching variables.
4. Specify ties = Breslow for $1:m$ matching or ties = discrete for $n:m$ matching. Ties are observations sharing identical event time values. In the current example, ties = discrete is used.

To relate the conditional logistic regression for $n:m$ matching to the hazard rate model, the procedure described above treats each matching set as a risk set, and controls as right-censored. So only one numerical value of the event time variable matters. Cases are event occurrences, and controls are nonoccurrences. Just as in a proportional-hazards model in which the parameters for stratifying variable(s) for the baseline hazard function are eliminated, those for the matching variable(s) are nuisance parameters and are also eliminated in the conditional logistic regression model.

Fitting a conditional logistic regression model to the data in the chapter appendix while using the matching variable age as the stratifying variable of the Cox's regression in SAS, we obtain the results in Table 7.2. No summary is presented of the numbers of events and censored values. Such a summary should be useful if the researcher is in doubt whether the analysis has been conducted correctly. The number of events in each stratum (risk set) should be identical to the number of cases in each matched set; the number of censored observations in each stratum should be identical to the number of controls in each matched set; the total number of strata should equal the total number of matched sets. Any discrepancy should alert the researcher to go back to the program to make sure that it is set up correctly according to the steps described above.

Table 7.2 Comparing Cases and Controls for the Data in Appendix 7.A

x-Variable	β	se(β)	Wald χ^2
WLM	−0.015	0.007	4.500
SDP	0.808	0.368	4.822
HT	1.751	0.739	5.612
UI	0.883	0.480	3.383
LRT, df = 4	17.961		
WT, df = 4	15.558		
LMT, df = 4	17.315		

The estimates in Table 7.2 give the comparison of the effect of certain risk factors between the cases and controls with regard to whether a LBW event will occur. The results show that the effect of weight at last menstruation on having a LBW baby is significantly smaller for women who actually experienced a LBW (cases) than for those who did not (controls). On the other hand, the effects of smoking during pregnancy and hypertension on having a LBW baby are significantly greater for cases than for controls. Uterine irritability shows no significant difference in effect between the case and control groups. Therefore, the comparison of β (effect on event occurring) plays out between the case and the control groups even though the cases experienced the event while the controls did not, because their information on having an event is censored. In theory, either group could experience the occurrence or nonoccurrence of an event.

7.3 DISPERSION HETEROGENEITY

The exponential family of distributions lies at the heart of the GLM. For the distributions described by (6.4), the dispersion parameter ϕ is assumed to be a constant, either directly estimable from the data or redundant because it is a simple function of μ. In addition to the link function $g(\cdot)$ discussed in Chapter 6, we may further define the relation between the variance of y and the mean by a variance function v (Breslow 1996, p. 23; McCullagh and Nelder 1989, p. 29),

$$\text{Var}(y) = \phi w^{-1} v(\mu), \qquad (7.3)$$

where the variance function $v(\cdot)$ depends on the canonical parameter [$v(\cdot) = b''(\theta)$] and hence the mean, and the w is a known *prior weight* that varies from observation to observation. For example, for a sample of observations, each of which is the mean of m independent readings or measurements, we have $w = m$. Thus, any variation in the variance must come through the canonical parameter (or the mean).

Two examples should help. For the normal distribution, $\phi = \sigma^2$ because the link is identity and $v(\cdot)$ is 1. For the binomial and Poisson models, $\phi = 1$ because $v(\mu) = \mu(1 - \mu)$ and $v(\mu) = \mu$, respectively.

There are two possibilities where the assumption about ϕ is violated:

- For the Poisson and binomial data (when $m > 1$), the dispersion may be greater (occasionally smaller) than unity. This is known as overdispersion (or underdispersion), and is typically generated by clustered data such as observations coming from the same units (families, litters, classes, schools, etc.). The measurements of the clusters are usually not included in the model.
- More generally, ϕ may be a function of some independent variables. A special case of the problem is heteroscedasticity in linear regression.

Put differently, overdispersion arises when the probability of an event occurring on a unit depends on the probabilities of events occurring on other units in the cluster and when that dependence is positive; underdispersion occurs when that dependence is negative. In the second and more general situation, statistical comparison in the GLM is affected, since ϕ may vary across groups, with some independent variables, or both. This is what we define as dispersion heterogeneity.

However, we should not confuse the correlation between the variances of **y** and **X** with that between the variances of **e** and **X**. It is a well-known fact that for GLMs the variance of **y** can be a function of the explanatory variables because the variance of **μ** is in the definition of the variance of **y** (7.3); in contrast, for the linear regression model homoscedasticity of the error variance guarantees that the variations in **μ** are not part of the error term. That is, the assumption of dispersion homogeneity is not violated if the variance of **y** varies only through **μ**. The assumption is violated if the errors systematically vary with another quantity apart from **μ**, such as the dispersion parameter ϕ varying with groups of observations.

For example, in linear regression where the link function is identity and $\phi = \sigma^2$, there is no particular nominal value for the variance ϕ, because ϕ can take on any positive numerical value as long as it is not a function of certain explanatory variables. Heteroscedasticity occurs when ϕ is a function of some x-variable. The same idea applies to other members of the GLM.

To correct for overdispersion (or underdispersion), we must first obtain an estimate of the dispersion parameter ϕ. For binomial and Poisson data, for example, overdispersion is often a problem; the dispersion parameter should be 1 (or $1/m$ when there is a binomial prior weight variable m). McCullagh and Nelder (1989) suggested estimating the dispersion parameter for GLMs with canonical links by

$$\hat{\phi} = \frac{1}{N - K} \sum_{i=1}^{N} \frac{(y_i - m_i \mu_i)^2}{m_i \mu_i (1 - \mu_i)} = \frac{\chi^2}{N - K}, \qquad (7.4)$$

where $m_i = 1$ for analyses without prior weights. Thus, the estimator is given by the residual sum of squares appropriately weighted by associated degrees of freedom. In other words, the ϕ estimate equals the Pearson deviance statistic divided by the df of $N - K$. The Pearson deviance, though it is often close to the (scaled) deviance for GLMs, is defined differently. The latter is defined by

$$2[\mathcal{L}(\mathbf{y};\mathbf{y}) - \mathcal{L}(\hat{\boldsymbol{\mu}};\mathbf{y})],$$

where $\mathcal{L}(\hat{\boldsymbol{\mu}};\mathbf{y})$ denotes the maximum of the log likelihood under the assumption that a given model holds, expressed in term of the $\boldsymbol{\mu}$ for an observed vector of $\mathbf{y}$. For a saturated model (i.e., a model which contains as many parameters to be estimated as independent observations), $\hat{\boldsymbol{\mu}} = \mathbf{y}$, and the (scaled) deviance equals zero. The number of degrees of freedom equals the difference between the number N of observations and the number K of parameters in the model. As the reader realizes, the (scaled) deviance has the same form as the LRT and is one case of it. The only difference is that the deviance has the reference model chosen to be the saturated model, while the LRT can use any model as the reference.

This method of estimation of ϕ described by (7.4) often yields a reasonable approximation (Breslow 1996; Smith and Heitjan 1993). The estimation is quite simple to calculate, because many computer programs produce the Pearson deviance statistic. Smith and Heitjan (1993) developed a score test for overdispersion in GLMs, and their procedure is implemented in an S-function available from Statlib on the Internet at Carnegie Mellon University. Even if a formal test is not conducted, a quick check of the deviance given its degrees of freedom should be done regularly.

Dealing with overdispersed clustered binomial or Poisson data (with no covariates or groups), Rao and Scott (1992, 1999) proposed a simple way to weight the data. To adjust an overdispersed model (with covariates), we multiply $\text{Var}(\hat{\boldsymbol{\beta}})$ by an estimate of the GLM scale factor, which is given by (7.4), or we weight the $\mathbf{y}$-data by such an estimate. Alternatively, for a GLM of Poisson data, we may use a negative binomial model to accommodate over-dispersion. To adjust dispersion in a GLM in comparing groups, we may adjust $\text{Var}(\hat{\boldsymbol{\beta}}_g)$ by estimates of ϕ_g. While we may always adjust the separate groups with $\hat{\phi}_g$ (since an adjustment with any $\phi_{\hat{g}}$ that is close to the nominal value will not change the standard errors much), we may want to have a significance test to determine whether any adjustment will be necessary. When comparing two groups g and h (for $g \neq h$), the Pearson χ^2 from each group may form an F-ratio based on (7.4) such that

$$F_{[N_g-K,\,N_h-K]} = \frac{\chi_g^2/(N_g - K)}{\chi_h^2/(N_h - K)}. \tag{7.5}$$

It is necessary to have the χ^2 for the group with greater dispersion in the numerator. More generally, for testing for significant departure from homogeneous dispersion among G groups, we may obtain a general F-test based on Pearson deviance measures as follows:

$$F_{[GK, N-GK]} = \frac{\sum_{g=1}^{G} |\chi_g^2 - (N_g/N)\chi^2|/GK}{\chi^2/(N-GK)}, \qquad (7.6)$$

where N and χ^2 pertain to the entire sample. The F-test is intended to detect dispersion heterogeneity regardless of overdispersion or underdispersion, and is flexible enough to accommodate multiple groups. A significant result indicates a violation of the assumption of homogeneous dispersion among the groups. This general F-test does not give conclusions identical to those drawn from the pairwise F-test of (7.5), which depends on which particular pair of groups chosen for comparison.

Recall that the LRT discussed as (6.15) assumes dispersion homogeneity. When the assumption is violated, the test is no longer a valid method for assessing parameter equality. Under heterogeneous conditions, we adjust the variance–covariance matrix with the dispersion factors. The likelihood functions after the adjustments are no longer true likelihood, but quasilikelihood functions, though their characteristics and behavior will be the same as for likelihood functions (McCullagh and Nelder 1989, Chapter 9). A straightforward extension of the GLM is a GLM based on quasilikelihood (Nelder 1998). For testing the hypothesis (6.11), the test (6.15) for comparing multiple groups under dispersion heterogeneity becomes a quasilikelihood ratio test,

$$\text{QLRT} = -2Q\mathcal{L}(\hat{\boldsymbol{\beta}}) - \sum_{g=1}^{G} \left[-2Q\mathcal{L}(\hat{\boldsymbol{\beta}}_g)\right] \sim \chi^2. \qquad (7.7)$$

Furthermore, we cannot compare the -2 quasilikelihood functions obtained from the (homogeneous) models after adjustments with the -2 likelihood function from the combined (heterogeneous) model without adjustment. The corresponding quasilikelihood functions adjusted by the $\hat{\phi}_g$ estimates for the pooled model, $Q\mathcal{L}(\hat{\boldsymbol{\beta}})$, can be obtained by weighting y_{gi} differently with $\{\hat{\phi}_g\}$ according to their membership in g, that is,

$$y_{gi}^{(w)} = \frac{y_{gi}}{\hat{\phi}_g} \qquad (7.8)$$

where $y_{gi}^{(w)}$ are the weighted y_i for each group g. For grouped data, both y_i (events) and n_i (trials) variables should be weighted. For individual data, weighting y_i means transforming it, thus changing its expected values, though the $\hat{\beta}_k$ are unaffected. This method of weighting is consistent with the general recommendations made by Clogg and Eliason (1987) about weighting in loglinear models. Just as sampling weights may apply to loglinear models (Clogg and Eliason 1987), the method discussed here should apply to data drawn from either a simple random sampling design or from a complex sampling design. The QLRT statistic (7.7) is formed in a way analogous to the LRT (6.15). Similarly, the Wald statistic and the Lagrange multiplier or score statistic based upon quasilikelihood can be calculated.

Under homogeneous conditions with $\phi_g = \phi$ = nominal dispersion (1 for the Poisson and the logit model, for example), the LRT and the QLRT (or the Wald test or the score test before and after the adjustment) will give identical conclusions, because the quasilikelihood functions will have the same values as the likelihood functions. Under dispersion heterogeneity, the three test statistics based on quasilikelihood assess the hypothesis of parameter equality between the groups where the dispersions are adjusted to be homogeneous.

Another approach to adjusting the variance of $\hat{\boldsymbol{\beta}}$ with estimated ϕ's uses the observed covariance matrix of the scores instead of the expected information for computing the asymptotic variance, thereby arriving at the so-called *sandwich* variance estimator

$$\mathrm{Var}(\hat{\boldsymbol{\beta}}) = \mathcal{I}^{-1} \left[\sum_{Li=1}^{N} w_i^2 \left(\frac{y_i - \mu_i}{v(\mu_i)} \right)^2 x_i x_i' \right] \mathcal{I}^{-1},$$

where $\mathcal{I}$ is the Fisher information matrix without the scale factor ϕ, μ_i is evaluated at the fitted value $\hat{\mu}_i$, and the middle factor in square brackets gives the observed covariance matrix of the scores (Breslow 1996). The sandwich estimator has been known as the White estimator or Huber estimator, depending on its form and context of application, and it has been implemented in major statistical software packages (e.g., the robust estimator in SAS and Stata, which also allows its use on clustered data). The sandwich variance estimator often seems to underestimate the true variability with moderate sized samples, and it may be preferable to employ a resampling method such as the bootstrap or the jackknife to estimate variance (Breslow 1996).

The first approach, of adjustment with ϕ_g, assumes that ϕ varies with g only. Although we may generalize the method to adjustment by more than the grouping variable, the estimation of individual ϕ's according to the levels of the variables under consideration and the subsequent weighting will quickly become intractable. The second approach, by way of the sandwich estimator, though it appears to be flexible, assumes that the dispersion parameter varies only with **X**, the variables responsible for $\boldsymbol{\mu}$. This may, but does not have to, be true. In addition, from a practical point of view, procedures that implement the sandwich estimator may not compute quasi-likelihood functions for making comparison. This makes an application of the QLRT one step more difficult, even though the new $\mathrm{Var}(\hat{\boldsymbol{\beta}})$ can be inserted into a calculation of the WT without further complication.

For dealing with the problem of dispersion heterogeneity more directly and more flexibly, there exists two related modeling approaches—double generalized linear models (DGLMs) (Smyth and Verbyla 1996, 1999) and mean and dispersion additive models (MADAMs) (Rigby and Stasinopoulos 1995, 1996). The two methods bear many similarities but differ in their generality. The DGLM specifies, in addition to the link function

$$g(\mu) = \eta = X\beta, \qquad (7.9)$$

another link function for the dispersion,

$$h(\phi) = \xi = Z\lambda, \qquad (7.10)$$

where $h(\cdot)$ can be different from $g(\cdot)$, Z is a matrix of covariates affecting ϕ, and λ contains their parameters.

DGLMs specify link functions for $h(\cdot)$ as GLMs do for $g(\cdot)$ in that both the mean and the dispersion link functions in the DGLM are parametric. The *generalized additive model* (GAM) extends the GLM by allowing the link function $g(\cdot)$ to be nonparametric, such as various smoothing functions (Hastie and Tibshirani 1990). In a similar fashion, the MADAM allows both $g(\cdot)$ and $h(\cdot)$ to be nonparametric. Therefore, the MADAM can be viewed as a general model for the GLM, the GAM, and the double GLM. Indeed, both the DGLM and the MADAM are extensions of the GLM along two of the major directions that Nelder (1998) identified. In this book, we focus on parametric links $h(\cdot)$ similar to the link $g(\cdot)$ discussed in Chapter 6.

The DGLM assumes that **y** follows an exponential dispersion model, a theoretically justified assumption. Estimation of the DGLM is via restricted ML. Given any working value for λ, the DGLM estimates β using an ordinary GLM for the y_i with weights w_i/ϕ_i. Similarly, given any working value for β, the DGLM estimates λ using a gamma GLM for the d_i, where $d_i = d(y_i, \mu_i)$ for the directed distance between y_i and μ_i (Smyth 1989). We call the GLM for estimating β for fixed λ the *mean submodel*, and the gamma GLM for estimating λ for fixed β the *dispersion submodel*; the latter has a dispersion parameter of 2. The estimation scheme works well by alternating between estimating β for fixed λ and λ for fixed β, because β and λ are orthogonal parameters (Smyth 1989). The DGLM is implemented in an S-function available at http://www.maths.uq.edu.au/~gks/s/dglm.html; the MADAM is implemented in a Glim macro available at http://tara.unl.ac.uk/~11stasinopou/madam.html.

In the example section we will illustrate the use of adjustment via ϕ_g and DGLM.

7.3.1 The Data

The data presented in Table 6.1 are actually aggregated over nine geographic regions or clusters. A potential clustering variable for causing dispersion heterogeneity in this case might be the precinct the respondent resided in. Lacking such information or even the knowledge of SSAs or SMSAs, the data

Table 7.3 1980 U.S. Presidential Vote among White Americans

Region	Sex	Political Views						
		1	2	3	4	5	6	7
New England	F	1/3	1/7	6/18	23/45	9/16	8/12	1/1
	M	1/1	0/9	3/14	16/32	12/21	10/14	1/1
Mid Atlantic	F	0/5	9/31	7/22	64/119	17/30	18/31	2/4
	M	1/4	1/16	9/22	34/63	35/50	24/29	4/7
E. N. Central	F	1/7	7/20	9/39	50/106	37/62	35/45	4/8
	M	0/8	4/26	18/46	63/109	42/68	31/42	6/9
W. N. Central	F	1/2	4/21	5/14	34/68	22/40	9/15	1/2
	M	0/4	5/12	2/12	20/44	20/33	19/22	1/2
S. Atlantic	F	1/5	7/25	8/24	42/92	24/48	27/39	2/5
	M	0/3	4/21	7/18	34/64	28/39	43/55	6/8
E. S. Central	F	0/0	2/9	3/8	13/36	8/19	15/24	5/5
	M	0/1	2/6	2/8	11/25	13/16	9/16	2/6
W. S. Central	F	0/0	3/10	7/20	19/40	10/19	17/22	4/5
	M	0/1	1/4	5/13	20/47	13/16	19/25	0/3
Mountain	F	0/0	1/5	4/14	22/41	11/14	14/16	4/5
	M	0/0	3/3	4/12	16/27	16/22	20/24	2/2
Pacific	F	1/4	7/29	7/19	40/79	23/33	29/37	3/4
	M	0/1	4/21	8/25	21/40	33/41	22/25	3/4

Source: 1982, 1983, 1984, and 1985 General Social Surveys.
Note: Political views range from 1 = extremely liberal to 7 = extremely conservative. Each entries contains two counts: (number of people who voted for Reagan)/(total turnout).

are presented by region of residence. Observations from the same cluster may behave more similarly, thus contributing to within-cluster correlations, which in turn may contribute to dispersion heterogeneity. Nonconstant dispersion may further affect the test of parameter equality between groups. To evaluate the complication of dispersion heterogeneity, we revisit the data in Table 6.1. The data there are now broken down by regions and given in Table 7.3. For illustrative purposes, we have only nine such clusters. In reality, the number of clusters may be much greater.

Notice that there are four observations in Table 7.3 with a zero count for the turn-out. These cases will be omitted from the analysis in the following subsection, because we use a logit link for modeling these data. Zero counts may not be a problem for another link function. Cases with a value of zero would not present a problem for an identity link, for example.

7.3.2 Group Comparison with Heterogeneous Dispersion

To assess the influence of dispersion heterogeneity on testing parameter equality, we consider three models. The first model is the conventional GLM. It is a naive model in that constant ϕ is assumed. To simplify comparison

Table 7.4 GLM and DGLM Estimates for Logit Models of the Voting Data

	GLM			DGLM			
$x \times z$ Variable	$\hat{\beta}$	se_1	se_2	$\hat{\beta}$	se	$\hat{\lambda}$	se
Intercept	−2.347	0.217	0.282	−2.494	0.253	−0.208	0.320
Female	0.384	0.290	0.329	0.502	0.295	−0.752	0.256
Political views	0.595	0.049	0.063	0.630	0.060	0.143	0.065
Female × pol. views	−0.132	0.066	0.074	−0.160	0.070	—	—
(Q)LRT for $\beta_m = \beta_f$	—	9.188	7.414		8.281		
df for (Q)LRT	—	2	2		2		

Note: The adjusted standard errors from the GLM are labeled se_1, and the standard errors from the GLM adjusted by a group-specific $\hat{\phi}_g$ estimate are labeled se_2. The (Q)LRT is for testing the parameter difference of the intercept and political views between males and females.

across the methods for dispersion heterogeneity, we use the one-model test with interactions involving the grouping variable. The second model assumes ϕ varies with the groups, and adjusts the standard errors using the weights $\hat{\phi}_g$ either as scale weights in fitting GLM or as data weights to be applied on the y-vectors. The third model relaxes the assumption further by modeling ϕ as a function of some z-variables using a log link. Table 7.4 presents the results from the three models and the (Q)LRTs for testing parameter equality between the groups based on the three models.

The first two models share the same parameter estimates, because only the standard errors are adjusted with $\hat{\phi}_g$. The $\hat{\phi}_g$ for the males is 1.6913, and that for the females is 0.7639, using (7.4); it is 1.2398 for the combined model including both sexes. This indicates a sizable amount of dispersion heterogeneity between the two groups, whether or not there exists overdispersion in the combined model. The F-ratio of (7.5) comparing the χ_g^2 of the males with that of the females gives a value of $2.214_{[60, 58]}$, significant at the 0.01 level. Using the general F-test of (7.6) for detecting dispersion heterogeneity, we obtain $11.038_{[4, 118]}$, also significant at the 0.01 level. Hence, either F-test suggests a violation of dispersion homogeneity between groups. The two $\hat{\phi}_g$ estimates are used as scaling weights for reestimating the two models. The parameter estimates remain unchanged by the correction, but the standard errors change. The adjusted standard errors are reported as se_2, next to the nominal standard errors without adjustment (se_1). However, no substantive conclusions are affected, because all the parameter estimates (except that for the sex variable) are large relative to their standard errors, be they nominal or adjusted. Overall, though, the adjusted standard errors are all larger than the unadjusted ones. After adjustment, dispersion parameter estimates are all unity, as they should be by assumption.

Due to dispersion heterogeneity, the LRT presented in Chapter 6 may not assess the hypothesis of parameter equality adequately. Thus, we turn to tests based on quasilikelihood. First, the quasilikelihood ratio test (7.7) can be performed by taking the difference between the residual deviance measure of the full model and that of the model without the two terms involving the sex variable. This results in a QLRT statistic of 7.414 with two degrees of freedom. The results suggest that, were women and men to have identical dispersion, the difference between the parameter estimates of the two groups would be less significant than before the adjustment, though still significant at the 0.05 level.

The four columns under the heading of DGLM in Table 7.4 present the β and λ estimates and their respective standard errors for the mean and the dispersion submodels. It appears that the standard errors for the β's are all larger than those from the naive model assuming dispersion homogeneity, but smaller than those from the model weighted with $\hat{\phi}_g$. In this DGLM, ϕ is assumed to be a function of two z-variables, sex and political views. The ratios of λ to standard errors indicate that both variables are responsible for the heterogeneous dispersion. Because the vector $\boldsymbol{\beta}$ is estimated iteratively with updated $\boldsymbol{\lambda}$ values, the final estimates in the vector are different from those from the GLM, with or without the adjustment with $\hat{\phi}_g$.

The QLRT statistic for testing the hypothesis of parameter equality between the sexes, $\boldsymbol{\beta}_m = \boldsymbol{\beta}_f$, is obtained by fitting two DGLMs. The first is a DGLM with only one explanatory variable, political views, in the mean submodel while keeping both sex and political views in the dispersion submodel. The second DGLM contains the same dispersion submodel specification, but with both the sex main effect and its interaction with political views back in the mean submodel as presented in Table 7.4. The difference between the residual deviance measures of these models gives a test of (7.7). Just like the standard errors in the mean submodel of the DGLM, the QLRT statistic for the DGLM also falls between those from the naive model and the GLM adjusted with $\hat{\phi}_g$.

In studying parameter comparisons, we have found that dispersion heterogeneity not only can affect the standard errors and thus the test statistic of parameter equality, but also can change the β-estimates in the mean submodel. Thus, we ought to be prepared for dispersion heterogeneity even when our purpose in modeling is not to compare across groups. A five-step procedure may help organize the task at hand:

1. Calculate a ϕ for the original, pooled model. A departure from the nominal value in the case of the binomial and Poisson data should warrant further exploration. A seemingly nominal ϕ, however, may not mean homogeneity, because variations across groups may cancel one another out for the overall ϕ-estimate.
2. Determine which z-variables may be responsible for a varying ϕ. When in doubt, try out suspected variables before trimming the dispersion submodel.

3. Test for equality in β by fitting two DGLMs with and without the grouping variable and its interactions with the x-variables of interest in the mean submodel, while keeping the same dispersion submodel.
4. If equality in λ among the groups becomes a hypothesis of interest, then fit two DGLMs with and without the grouping variable and its interactions with the z-variables of interest in the dispersion submodel, while keeping the same mean submodel.
5. If the hypotheses of equality in both β and λ are of interest, then fit two DGLMs with and without the grouping variable and its interactions with the x-variables of interest in the mean submodel, and with and without the grouping variable and its interactions with the z-variables of interest in the dispersion submodel. Keep in mind that a significant statistic in a joint test like this may be due to group differences in one of the two submodels. This suggests that step 5 must be carried out together with steps 3 and 4 to identify the source of inequality among the groups.

While the example above only illustrates testing equality in β in the first three steps, the extension to testing equality in λ among groups of observations should be straightforward, especially with the ready availability of the DGLM function in S-plus.

7.4 BAYESIAN GENERALIZED LINEAR MODELS

In recent years Bayesian statistics has developed from a marginalized methodology to a pillar in contemporary statistics, applied and theoretical. The GLM is not immune to this development. Because the Bayesian approach is a general statistical method, there are many ways in which the GLM can be modified, as reflected by the contributions in *Generalized Linear Models: A Bayesian Perspective* (Dey, Ghosh, and Mallick 2000). In this section we focus on one method in which Bayesian thinking may benefit group comparison in GLMs. First, however, let us review some of the fundamentals of Bayesian statistics.

7.4.1 Bayesian Inference

Bayesian inference allows the researcher to bring subjectivity or belief (prior probability) into calculating probability and performing inference (posterior probability) in the light of the data (via the likelihood function). Let $P(H \mid B)$ denote the subjective or believed probability of an event occurring or of observing the data, that is, the probability of a hypothesis H given prior information or belief; this is known as the *prior probability*. Let $P(D \mid H, B)$ denote the *likelihood*, which gives the probability of the data assuming that the hypothesis H and the prior information or belief B are true. Let $P(H \mid D, B)$ denote the probability of H after considering the data and prior knowledge; this is known as the *posterior* probability. The Bayes rule or theorem is given by

$$P(H|D, B) = \frac{P(H|B)P(D|H, B)}{P(D|B)},$$

where $P(D \mid B)$, the probability of the data given prior knowledge, is regarded as a normalizing or scaling constant, because it is unrelated to the hypothesis or model under consideration. Thus, the posterior probability of a hypothesis is proportional to the prior probability times the likelihood.

All the probabilities in the above equation are conditional. We may also have unconditional probabilities. Let H_p denote all exhaustive and mutually exclusive hypotheses regarding the event or data, and let us drop B from the probabilities, since it is understood. Then

$$P(D) = \sum_p P(D|H_p)P(H_p).$$

This is known as the generalized addition law. Another law is the generalized multiplication law

$$P(H_1 H_2 \ldots H_P) = P(H_1)P(H_2|H_1)P(H_3|H_1 H_2) \cdots P(H_P|H_1 H_2 \ldots H_{P-1}).$$

The key result deduced from the generalized addition law is the Bayes theorem,

$$P(H_p|D)P(D) = P(D, H_p) = P(H_p)P(D|H_p),$$

and provided that $P(D) \neq 0$,

$$P(H_p|D) \propto P(H_p)P(D|H_p). \quad (7.11)$$

This is one of several ways of stating the theorem and is a simplified version of the formula including $P(D)$ as a scaling constant in the denominator.

When H_p indicates all mutually exclusive and exhaustive hypotheses 1 to P, the Bayes theorem leads to

$$P(H_p|D) = \frac{P(H_p)P(D|H_p)}{\sum_{p=1}^{P} P(H_p)P(D|H_p)}. \quad (7.12)$$

7.4.1.1 An Example
As an example to illustrate a simple use of the Bayes theorem, let us look at the political situation surrounding people's attitudes toward to the approval of the abortion pill in the U.S. In September 2000, the U.S. Food and Drug

Administration (FDA) approved for sale of the French abortion pill RU-486. A typical Democratic stand on legal abortion, as voiced by the Democratic presidential candidate Al Gore, was yes, while a typical Republican stand on the issue was no except for extreme circumstances, as expressed by the Republican presidential candidate George W. Bush. The candidates' views were made clear during debates leading up to the November 2000 general election. Suppose that the Democrats (DM) won the election with 53% of the popular support while the Republicans (RP) garnered 47% (assuming for simplicity support for all other candidates was negligible). Suppose again that following the general election a public opinion poll showed that 54% of the Republican voters and 86% of Democratic voters said yes (Y) to the Clinton administration's decision to legalize RU-486 in the U.S. Finally, suppose that you met someone at the time who said yes to the survey question about the FDA's approval of RU-486. You are, however, interested in guessing which political party this person supported. Given all this information and following the Bayes theorem, we have

$$P(DM|Y) = \frac{P(Y|DM)P(DM)}{P(Y|DM)P(DM) + P(Y|RP)P(RP)}$$

$$= \frac{(0.86)(0.53)}{(0.86)(0.53) + (0.54)(0.47)} = 0.64.$$

Thus, the probability of this person having supported the Democratic Party is 0.41.

7.4.1.2 Bayesian Model Comparison

Suppose we have two competing hypotheses, H_p, for $p = 0, 1$. As before, H_0 represents $\beta_1 = \beta_2$ for two groups of observations. From the Bayes theorem (7.12), we obtain

$$P(H_p|D) = \frac{P(D|H_p)P(H_p)}{P(D|H_0)P(H_0) + P(D|H_1)P(H_1)} \quad \text{for } p = 0,1 \quad (7.13)$$

where the datum D is assumed to have arisen under one of the two competing hypotheses H_0 and H_1 with probability densities $P(D|H_0)$ and $P(D|H_1)$, and $P(H_p|D)$ gives the posterior probability of H_p given data. The ratio of the posterior probability of H_1 to that of H_0 gives the posterior odds, measuring the degree to which the data support H_1 versus H_0:

$$\frac{P(H_1|D)}{P(H_0|D)} = \frac{P(D|H_1)}{P(D|H_0)} \frac{P(H_1)}{P(H_0)}, \quad (7.14)$$

where the first factor on the right is the Bayes factor, and the second gives the prior odds. Unlike the abortion pill example in Section 7.1.4.1, often we do not have a priori knowledge to favor one hypothesis over another, and so assign an equal value to each prior probability. With two competing hypotheses, we have $0.5/0.5 = 1$ for the prior odds.

To obtain posterior odds of comparing hypotheses in (7.14), one only needs likelihood ratios, and Raftery (1996) defined the likelihoods appearing in (7.13) and (7.14) to be

$$P(D|H_p) = \int P(D|\theta_p, H_p) P(\theta_p|H_p) d\theta_p,$$

where θ_p designates some parameters of interest for hypothesis p. Raftery (1996) proposed to use the Laplace method for integrals to approximate the likelihood defined here. There are three methods of approximation of the Bayes factor (Raftery 1996), of which the most accurate and the simplest are presented here.

The most accurate approximation relies on the Laplace method and expresses twice the natural logarithm of the Bayes factor as

$$2\ln B_{10} \approx \chi_{10}^2 + (E_1 - E_0), \qquad (7.15)$$

where B_{10} denotes the Bayes factor contrasting H_1 and H_0, with the model representing H_0 nested within that representing H_1, and χ_{10}^2 is the difference between the likelihood ratios χ^2 from the models representing H_1 and H_0, respectively. The E_p (where $p = 1, 0$) gives a quadratic measure of how well model p (the model under hypothesis p) performs in terms of the closeness of $\hat{\theta}_p$ to $E(\theta_p|H_p)$. For computational details, see (Raftery 1996, p. 254).

The simplest approximation is the *Bayesian information criterion* (BIC) a direct and simple adjustment to the LRT:

$$2\ln B_{1-0} \approx \chi_{1-0}^2 - df_{1-0} \ln N, \qquad (7.16)$$

where df_{1-0} is the difference in degrees of freedom for the models under hypothesis 1 and hypothesis 0. H_0 can either be a baseline model such as the saturated model, or the null model that constrains the parameters of the independent variables. When the baseline is the saturated model, the LRT used is a *deviance* measure. When the baseline model is the null (i.e., the model without any independent variables), the formula for the BIC becomes

$$2\ln B_{1-0} \approx -\chi_{1-0}^2 + df_{1-0} \ln N.$$

This maintains the consistency of the BIC in model comparisons. The statistic given by (7.16) is also known as Schwarz's criterion. BICs can also be applied to models that are not nested within each other.

However, testing parameter equality across groups involves more than contrasting H_0 and H_1, which may simply represent two models in the space of all possible models. In this chapter we consider the method of Bayesian model comparison among multiple groups as given in Liao (2002). For example, for the null hypothesis $\boldsymbol{\beta}_1 = \boldsymbol{\beta}_2$ with each vector $\boldsymbol{\beta}_p$ containing two parameters β_{p1} and β_{p2}, the standard alternative hypothesis $\boldsymbol{\beta}_1 \neq \boldsymbol{\beta}_2$ actually contains the three subhypotheses $\beta_{11} \neq \beta_{21}$, $\beta_{12} \neq \beta_{22}$, and $\beta_{11} \neq \beta_{21}$ and $\beta_{21} \neq \beta_{22}$ jointly true. This means that there are *four* models in total in the model space, one representing H_0 and three representing H_1.

This simple example suggests that, when comparing across groups, we need to allow for model uncertainty by considering all acceptable models in the relevant model space, and define and compute the posterior probabilities accordingly. Suppose that P models, $H_0, H_1, \ldots, H_P$, are being considered. Each of the hypotheses is tested against H_0, a baseline hypothesis, yielding Bayes factors $B_{1\text{-}0}, B_{2\text{-}0}, \ldots, B_{P\text{-}0}$ (Kass and Raftery 1995). The posterior probability of H_p when all the relevant models are considered is

$$P(H_p|D) = \frac{\alpha_p B_{p\text{-}0}}{\sum_{p=1}^{P} \alpha_p B_{p\text{-}0}}, \qquad (7.17)$$

where $\alpha = P(H_p)/P(H_0)$ is the prior odds of supporting H_p instead of H_0. A natural choice is to take all prior odds as 1. We may reexpress (7.17) as function of $2\ln B_{p\text{-}0}$ (Raftery 1995):

$$P(H_p|D) \approx \frac{e^{-0.5\ln B_{p\text{-}0}}}{\sum_{p=1}^{P} e^{-0.5\ln B_{p\text{-}0}}}. \qquad (7.18)$$

The summation is over all relevant models in the model space for testing a particular set of parameter equality restraints across groups under consideration. When testing parameter equality across multiple groups, we should include all the hierarchical (i.e., nested) models involving the parameters of interest in the model space.

With the assistance of (7.18) we may formulate quite flexible hypotheses for testing parameter equality across groups. This means either the null hypothesis or the alternative hypothesis or both may be a composite hypothesis represented by multiple models. Using again our simple example of testing the equality of two parameters between two groups, we may be interested in seeing as the alternative hypothesis parameter inequality involving at least one pair of nonintercept parameters. That is, our alternative hypothesis, H_1, consists of two models:

- $\beta_{10} = \beta_{20}$ but $\beta_{11} \neq \beta_{21}$;
- $\beta_{10} \neq \beta_{20}$ and $\beta_{11} \neq \beta_{21}$.

Our null hypothesis H_0, on the other hand, consists of two models as well:

- $\beta_{10} \neq \beta_{20}$ but $\beta_{11} = \beta_{21}$;
- $\beta_{11} = \beta_{21}$ and $\beta_{12} = \beta_{22}$.

Thus, out of the total of four models, two models form the null hypotheses, and two models form the alternative hypothesis.

To summarize: because of the flexibility of the Bayesian approach, we may test three useful types of hypotheses (H_1) of parameter equality across groups. Here we use H to indicate a general hypothesis and M_k to denote the models representing a hypothesis. One way for testing equality is through using interaction terms between the group variable and the explanatory variables. We use this approach to facilitate testing in the generalized linear Bayesian setting:

1. **Hypothesis type A (H_A):** A conventional hypothesis stating that all model parameters are different across all groups jointly. This is the standard test conducted with LRT, BIC, and other equivalents when only two models are compared. If we use the method of dummy variable with interactions, the null hypothesis states that only the main effects of the explanatory variables (i.e., no grouping variables) are in the model.

2. **Hypothesis type B (H_B):** In the set of all possible models, a subset of models that are nested within the remaining models are considered together as the null hypothesis, and the remaining models together represent the alternative hypothesis. The example in the next section, of comparing two parameters between two groups, is such a case.

3. **Hypothesis type C (H_C):** This is a generalization of type B. Now the models representing the null hypothesis (H_0) of parameter equality across groups are not necessarily nested within those representing the alternative hypothesis (H_1).

For testing a conventional hypothesis such as H_A, we consider two models: H_0 is represented by the model with the main effects of the explanatory variables only, while H_1 is represented by the model with all the pairwise interactions between the grouping variable(s) and the explanatory variables. When testing hypotheses such as H_B, we consider all models with any possible interactions between any group and any explanatory variables as representing H_1, and all models without such interactions in the model space as representing H_0. H_C is defined ad hoc by the modeler, and may not be nested.

For assessing these types of hypotheses, we obtain the follow formula for constructing posterior odds:

$$\frac{P(H_\Omega|D)}{P(H_{\bar{\Omega}}|D)} = \frac{\sum_{M_k \in \Omega} \alpha_k(M_\Omega)}{\sum_{M_k \notin \Omega} \alpha_k(M_{\bar{\Omega}})} \frac{\sum_{M_k \in \Omega} P(M_k|D)}{\sum_{M_k \notin \Omega} P(M_k|D)}, \qquad (7.19)$$

where Ω is the set of all models representing the alternative hypothesis H_Ω (which can be H_A, H_B, or H_C), $\alpha_k(M_\Omega)$ is the prior probability for model k, $\alpha_k(M_\Omega)/\alpha_k(M_{\bar{\Omega}})$ is the prior odds for model k that supports H_Ω versus $H_{\bar{\Omega}}$, and the summed (posterior) probability ratio (the second factor on the right-hand side) is adjusted by its prior odds, which in practice (when all prior probabilities are equal) is the ratio of one minus the number of models in Ω to the total number of models. For example, for three independent groups (two unique group contrasts) with two independent variables in the model, there are 47 models in the model space, 31 of which represent an H_B-type hypothesis of H_Ω while the remaining 16 represent $H_{\bar{\Omega}}$. Having 31 models is likely to produce an overrepresentation due to summing the 31 posterior probabilities as opposed to summing the 16 posterior probabilities of the non-Ω models. Thus, both posterior probabilities and posterior odds must be weighted according to their prior probabilities, in this case their likelihood of being represented in the model space. Similarly, for testing H_C, we need to sum the posterior probabilities of C_1 in the numerator and the posterior probabilities of C_0 in the denominator, and adjust with their prior odds accordingly.

7.4.1.3 The Model Space

For testing parameter equality among multiple groups, we include in the model space all models that satisfy the hierarchical principle (i.e., interactions are included only when their related main effects and lower-order interactions are present). Thus, the model space contains all models, starting from the one with the intercept only, and proceeding to the ones with all possible combinations of group and explanatory variables, and the ones with all possible combinations of interactions between the grouping variable(s) and the explanatory variables, as long as the hierarchical condition is met.

Following this logic, the model space includes two sets of permuted models satisfying the hierarchical principle with and without the interactions, and the total number of models (NM) is given as

$$\text{NM} = \sum_{r=0}^{G+J} \binom{G+J}{r} + \sum_{s=1}^{G} \sum_{t=1}^{J} \left[\binom{G}{s} \binom{J}{t} \sum_{u=1}^{s \times t} \binom{s \times t}{u} \right]$$

$$+ \sum_{r=1}^{G \times J} \binom{G \times J}{r} \quad \text{for } s \times t < G \times J, \qquad (7.20)$$

where $\binom{n}{r}$ is the number of unordered distinct combinations selecting r objects out of n (i.e., $n!/[(n-r)!r!]$), G is the total number of unique group contrasts, and J is the total number of explanatory variables. For example, for a situation with two unique group contrasts and two independent variables, there exist a total of 47 models in the model space.

A quantity of interest is the number of unique group contrasts in the model space. As with constructing unique contrasts between the multiple categories of a nominal variable, for G groups there exist only $G - 1$ unique contrasts. When all unique contrasts in the form of dummy variables representing a categorical variable are included in the model, the choice of the reference category should not affect an overall statistical test (such as the F or the χ^2) of the effect of the categorical variable. Similarly, the choice of the reference group should not affect the overall test of parameter differences among multiple groups as long as $G - 1$ contrasts are present. However, as with categorical variables, the choice of the reference category or group determines which categories or groups are represented by parameters, thereby influencing the interpretation. If a researcher is interested in testing group differences and in interpreting estimates, then for ease in interpretation use as the reference the group that has (close to) zero effects for the explanatory variables, so that other groups will be a deviation from this relative norm of zero effects.

7.4.1.4 Software

The software for estimating Bayesian GLMs with a single-step Newton approximation has been implemented by Adrian Raftery and his colleagues as an S-function (glib) available from the StatLib on the Internet at Carnegie Mellon University. The extension of the glib framework to comparison of multiple groups is implemented by the current author as an S function called multigroup.glib from the following web site:
http://www.staff.uiuc.edu/~f-liao/StatCompare/multigroup.glib

The program takes the same input as that for glib, with the addition of two variables, G (the number of group contrasts) and K (the number of independent variables, to be consistent with the indexing in glib, which reserves j for something else), and the output includes all the usual glib output plus posterior probability and posterior odds estimates for H_A and H_B. The testing of H_C is ad hoc and user defined. Once posterior probabilities are obtained for all the models in the model space from the output, the user can further construct relevant posterior probabilities and odds for testing H_C.

Alternatively, the data analyst may compute BIC based on LRT given by standard statistical software packages. This is feasible as long as the total number of variables is small enough to avoid generating too many models in the model space.

Table 7.5 Self-Esteem and Academic Achievement among College Students (Frequencies)

		Black		White	
Sex	GPA	High Self-Esteem	Low Self-Esteem	High Self-Esteem	Low Self-Esteem
M	High	15	9	17	10
	Low	26	17	22	26
F	High	13	22	22	32
	Low	24	23	3	17

Source: Demo and Parker (1987).

7.4.2 An Example

In this section we consider an example from Liao (2002) for testing for parameter differences in more than one independent variable between two groups. The substantive consideration is the relation between self-esteem and academic achievement. Do academic achievement and gender have an effect on self-esteem? If so, do these effects differ between black and white college students? These are questions that can be addressed by the data from Demo and Parker (1987). The grouping variable is race. This choice, though arbitrary, is substantively defensible and pedagogically useful. The data are presented in Table 7.5. Because the outcome is binomial, a logit link is used. Thus, two GLMs with the logit link, with one representing the null model and the other representing the full model (stratified by race), are fitted to the data. Thus, the model pooled over both races is

$$\ln\left[\frac{P(y_i = 1)}{1 - P(y_i = 1)}\right] = \beta_0 + \beta_1 \text{Sex}_i + \beta_2 \text{GPA}_i,$$

and the same model can be applied separately to the black and white observations to form the unrestricted model. Alternatively, the unrestricted model can take the form of dummy variables (representing the groups) with interactions:

$$\ln\left[\frac{P(y_i = 1)}{1 - P(y_i = 1)}\right] = \beta_0 + \beta_1 \text{Race}_i + \beta_2 \text{Sex}_i + \beta_3 \text{GPA}_i$$
$$+ \beta_4 \text{Race}_i \times \text{Sex}_i + \beta_5 \text{Race}_i \times \text{GPA}_i.$$

These two models constitute a conventional type A hypothesis of parameter equality between groups, with the first model representing the null hypothesis and the second model, with interactions involving the race variable, representing the alternative hypothesis. The ML estimates of these two models are presented in Table 7.6.

BAYESIAN GENERALIZED LINEAR MODELS

Table 7.6 Maximum Likelihood Estimates using the Self-Esteem Data

x-Variable	Pooled	Unrestricted	
		Black	White
Intercept	0.195	0.560	−0.250
	(0.190)	(0.281)	(0.271)
Sex (female = 1)	−0.708	−0.635	−1.141
	(0.242)	(0.336)	(0.389)
GPA (high = 1)	0.169	−0.287	0.941
	(0.242)	(0.341)	(0.391)
LRT: model χ^2	8.730	18.432	
Degrees of freedom	3	6	
BIC	−8.362	−15.751	
2 log B_{10} 1-step Newton:			
$\phi = 1$	−2.786	−7.927	
$\phi = 1.65$	−4.766	−12.821	
$\phi = 5$	−9.189	−23.848	

Note: The values in parentheses are standard errors of the parameter estimates. The three values of ϕ give the range of reasonable values of the standard deviation of the prior distribution.

The model in the "Pooled" column, the restricted model, assumes no difference in the effects of sex and GPA between the racial groups, while the model in the last two columns together is the unrestricted model with effects of x specific for each race. The LRT of either the model assuming parameter equality between the black and white students or the model assuming parameter inequality gives a significant χ^2 statistic at any conventional α-level. The difference of the two LRTs gives another LRT of whether the two groups are different. It appears they are significantly different with three degrees of freedom in the additional parameters estimated. According to the BIC difference of about 7.4, the unrestricted model of group differences is also preferred with moderately strong support.

For fitting Bayesian GLMs, the parameter ϕ represents the range of acceptable values from the reference set of proper priors. Specifically, in this case ϕ acceptably ranges from 1 to 5 in terms of the standard deviation of the prior distribution, with $\sqrt{e} \approx 1.65$ as the midrange point (Raftery 1996). At all levels of ϕ, the $2 \ln B_{10}$ estimates using the more exact single-step Newton method all likewise indicate that the restricted model fits the data better. The BIC values appear to be located somewhat higher in absolute value than the single-step Newton estimates when $\phi = 1.65$. The difference in BIC between the two models approximates the difference in $2 \ln B_{10}$ for the single-step Newton between the two models when $\phi = 1.65$. The two models, however, only represent two of the 13 possible models in the model space. Table 7.7 contains all models in the model space for the self-esteem data.

Table 7.7 Design Matrix for the Models of the Self-Esteem Data

Model	Race	Sex	GPA	Race × Sex	Race × GPA
1	0	0	0	0	0
2	1	0	0	0	0
3	0	1	0	0	0
4	0	0	1	0	0
5	1	1	0	0	0
6	1	0	1	0	0
7	0	1	1	0	0
8	1	1	1	0	0
9	1	1	0	1	0
10	1	0	1	0	1
11	1	1	1	1	0
12	1	1	1	0	1
13	1	1	1	1	1

The thirteen models in the model space ($P = 13$) are all possible models in the hierarchy for testing parameter equality between the two racial groups using the self-esteem data. From Table 7.7 we see that the conventional hypotheses of type A tested and presented in Table 7.6 are only two (models 7 and 13) of the thirteen possible models. If these are the only models in the model space, we may compute their respective posterior probabilities using (7.18). This results in a posterior probability of 0.9825 for the unrestricted model and 0.0175 for the restricted model, with posterior odds of 56 strongly against the model assuming equality of parameters. Kass and Raftery (1995) gave some guidelines for interpreting posterior probabilities and odds, which are summarized in Table 7.8.

In addition to the conventional hypothesis H_A, let us entertain hypotheses H_B and H_C. The first one, H_B, is a composite hypothesis that either the sex effects or the GPA effects or both are different between the two racial groups, the null hypothesis maintaining no such difference. The null hypothesis corresponding to the second, H_C, is another composite hypothesis that GPA has *no* effect, either as main effect or involved in any model with interaction with the grouping variable.

Table 7.8 Rules for Interpretation and Decision Making

B_{1-0}	$2 \ln B_{1-0}$	Evidence for H_1
< 1	< 0	Negative (supporting H_0)
1 to 3	0 to 2.2	Not worth more than a bare mention
3 to 12	2.2 to 5	Positive
12 to 150	5 to 10	Strong
> 150	> 10	Decisive

Source: Kass and Raftery (1995).

For testing H_B, the posterior probabilities of the last five models in Table 7.7 are summed for a single posterior probability representing H_B, which is compared against the posterior probability summed from the remaining models representing the null hypothesis. Both posterior probabilities and the resulting odds are adjusted accordingly by their prior odds using (7.19). Namely, the 5 to 8 ratio and the posterior probabilities are also rescaled to sum to 1. Put differently, the summed posterior probabilities representing the five models are weighted by multiplying by $1 - \frac{5}{13}$ while the summed posterior probabilities representing the eight models are weighted by multiplying by $1 - \frac{8}{13}$. The posterior probabilities are calculated similarly for H_C by summing over the posterior probabilities of the models involving the GPA variable (models 4, 6–8, 10–13) and those of the models not involving it (models 1–3, 5, and 9) and weighted accordingly.

Let us use a simple example to illustrate the weighting adjustment and the rescaling necessary. Suppose that we have ten models, three of which representing a null and seven of which representing an alternative hypothesis. Suppose again that each of them has an equal probability of 0.10. Without weighting, the total posterior null probability would be 0.30 and that for the alternative hypothesis would be 0.70; they are unequal because of the unequal number of models for each hypothesis. If we weight 0.30 by $\frac{7}{10}$ and 0.70 by $\frac{3}{10}$, then we have the null probability = 0.21 and the alternative probability = 0.21, representing equal posterior probabilities again: the unequal numbers of models do not affect the final posterior probabilities. Then the resulting probabilities can be normed or rescaled to sum up to 1, that is, $0.21/(0.21 + 0.21) = 0.50$, and so on. If for some reason one wanted the number of models represented, then one would choose not to weight the posterior probabilities.

The posterior probabilities and odds for hypotheses H_B and H_C as well as those for the conventional hypothesis H_A are presented in Table 7.9. The posterior probabilities and odds in Table 7.9 are for the alternative hypotheses, that is, the existence of actual differences between the groups. Judging

Table 7.9 Bayesian Posterior Information for the Data in Table 7.5

	Posterior Information		
	$\phi = 1$	$\phi = 1.65$	$\phi = 5$
Probability:			
H_A	0.9289	0.9825	0.9993
H_B	0.8333	0.9214	0.9909
H_C	0.4208	0.5318	0.8458
Odds:			
H_A	13.0737	56.1124	1,524.4909
H_B	4.9981	11.7148	109.0996
H_C	0.7264	1.1360	5.4835

by the posterior probabilities and posterior odds, the hypothesis of difference between the racial groups, either in the form of the conventional hypothesis H_A or in the composite form of H_B, is supported. However, on considering all relevant models in the model space, H_B gives much more moderate evidence against the hypothesis of no group differences (with the posterior odds ranging from 5 to 109). H_C, which states that GPA matters, either as an explanatory variable or in its differential effect between the racial groups, receives weak support from the data (with posterior odds ranging from 0.7 to 5.5).

We must realize that the example only shows how the Bayesian framework helps in capturing model uncertainty by considering the entire model space when making group comparisons. The flexibility of assigning prior probabilities has not yet been explored in the current example except by using proportional weights calculated for a model in the model space, but inclusion of reasonable prior knowledge is readily feasible by using user-assigned prior weights in the Bayesian GLM software discussed in this section.

7.A THE DATA FOR THE $n:m$ DESIGN

The data for the $n:m$ design discussed in Section 7.2.3 are presented in Table 7.10.

Table 7.10 Low Birth-Weight Data of an $n:m$ Design

Age	LBW	WLM	SDP	HT	UI	Age	LBW	WLM	SDP	HT	UI
16	1	130	0	0	0	18	1	148	0	0	0
	0	110	0	0	0		1	110	1	0	0
	0	112	0	0	0		0	107	1	0	1
	0	135	1	0	0		0	100	1	0	0
	0	135	1	0	0		0	100	1	0	0
	0	170	0	0	0		0	90	1	0	1
	0	95	0	0	0		0	90	1	0	1
17	1	130	1	0	1		0	229	0	0	0
	1	110	1	0	0		0	120	1	0	0
	1	120	1	0	0		0	120	0	0	0
	1	120	0	0	0	19	1	91	1	0	1
	1	142	0	1	0		1	102	0	0	0
	0	103	0	0	0		1	112	1	0	1
	0	122	1	0	0		0	182	0	0	1
	0	113	0	0	0		0	95	0	0	0
	0	113	0	0	0		0	150	0	0	0
	0	119	0	0	0		0	138	1	0	0
	0	119	0	0	0		0	189	0	0	0
	0	120	1	0	0		0	132	0	0	0

Table 7.10 (*Continued*)

Age	LBW	WLM	SDP	HT	UI	Age	LBW	WLM	SDP	HT	UI
19	0	115	0	0	0	23	1	97	0	0	1
	0	105	0	0	0		1	187	1	0	0
	0	235	1	1	0		1	187	1	0	0
	0	147	1	0	0		1	120	0	0	0
	0	147	1	0	0		1	110	1	0	0
	0	184	1	1	0		1	94	1	0	0
	0	120	1	0	0		0	130	0	0	0
20	1	150	1	0	0		0	128	0	0	0
	1	125	0	0	1		0	119	0	0	0
	0	120	0	0	0		0	115	1	0	0
	1	80	1	0	1		0	190	0	0	0
	1	109	0	0	0		0	123	0	0	0
	1	121	1	0	1		0	130	0	0	0
	1	122	1	0	0		0	110	0	0	0
	1	105	0	0	0	24	1	128	0	0	0
	0	105	1	0	0		1	132	0	1	0
	0	120	0	0	1		1	155	1	0	0
	0	103	0	0	0		1	138	0	0	0
	0	169	0	0	1		1	105	1	0	0
	0	141	0	0	1		0	90	1	0	0
	0	121	1	0	0		0	115	0	0	0
	0	127	0	0	0		0	110	0	0	0
	0	120	0	0	0		0	115	0	0	0
	0	170	1	0	0		0	133	0	0	0
	0	158	0	0	0		0	110	0	0	0
21	1	165	1	1	0		0	110	0	0	0
	1	200	0	0	1		0	116	0	0	0
	1	103	0	0	0	25	1	105	0	1	0
	1	100	0	0	0		1	85	0	0	1
	1	130	1	1	0		1	115	0	0	0
	0	108	1	0	1		1	92	1	0	0
	0	124	0	0	0		1	89	0	0	0
	0	185	1	0	0		1	105	0	0	0
	0	160	0	0	0		0	118	1	0	0
	0	110	1	0	1		0	120	0	0	1
	0	134	0	0	0		0	155	0	0	0
	0	115	0	0	0		0	125	0	0	0
22	1	130	1	0	1		0	140	0	0	0
	1	130	1	0	0		0	95	1	0	1
	0	118	0	0	0		0	241	0	1	0
	0	95	0	1	0		0	120	0	0	0
	0	85	1	0	0		0	130	0	0	0
	0	120	0	1	0	26	1	117	1	0	0
	0	130	1	0	0		1	96	0	0	0
	0	158	0	0	0		1	154	0	1	0
	1	112	1	0	0		0	190	1	0	0
	0	131	0	0	0		0	113	1	0	0
	0	125	0	0	0		0	168	1	0	0
	0	169	0	0	0		0	133	1	0	0
	0	129	0	0	0		0	160	0	0	0

Table 7.10 (*Continued*)

Age	LBW	WLM	SDP	HT	UI	Age	LBW	WLM	SDP	HT	UI
27	1	150	0	0	0	29	1	142	1	0	0
	1	130	0	0	1		0	107	0	0	1
	0	124	1	0	0		0	95	1	0	0
28	1	120	1	0	1		0	153	0	0	0
	1	95	1	0	0		0	110	0	0	0
	0	120	1	0	0		0	137	0	0	0
	0	120	0	0	0		0	112	0	0	0
	0	167	0	0	0	31	1	102	1	0	0
	0	140	0	0	0		0	100	0	0	1
	0	250	1	0	0		0	215	1	0	0
	0	130	0	0	0		0	150	1	0	0
29	1	130	0	0	1		0	120	0	0	0
	0	123	1	0	0	32	1	105	1	0	0
	0	150	0	0	0		0	121	0	0	0
	0	140	1	0	0		0	132	0	0	0
	0	135	0	0	0	32	0	134	1	0	0
	0	154	0	0	0		0	170	0	0	0
	0	130	1	0	0		0	186	0	0	0

CHAPTER 8

Comparison in Structural Equation Modeling

8.1 INTRODUCTION

As a generalization of linear regression analysis, structural equation modeling (SEM) has been increasingly applied in the social and behavioral sciences. There are at least two reasons for the popularity of SEM. First and foremost, SEM allows the data analyst to test, verify, and construct theories much more easily than conventional methods such as linear regression and factor analysis. The primacy of theory requires the empirical model to represent some theoretical model. Linear models rely on measured variables only, while factor analysis has as its purpose exploring the relationship between concepts and observed variables but not among concepts that are not directly measurable. SEM facilitates the representation of substantive theory with latent variables, which in turn are indicated by observed variables. Second, the availability of special-purpose software has greatly increased the applicability of SEM. With easy access to software such as LISREL, EQS, Mplus, and Amos, all on the Windows platform, the application of SEM is no longer confined to specialists.

Two types of variables are included in SEM: unobserved latent variables and observed indicator variables. Each latent variable, measured only in terms of observed variables, represents a substantive concept, and has an error term. By including an error, SEM recognizes the imperfection of our measurement of concepts. SEM also has the flexibility of including multiple equations in a single model via simultaneous estimation. This further extends the possible complexity of the model structure, and hence the elaborateness of the theory.

Latent variables can be continuous or discrete. SEM specifies latent variables to be continuous, even though the observed variables can be nonmetric. We consider discrete or categorical latent variables in the next chapter. In the following section we introduce the statistical background of

SEM. Readers unfamiliar with the topic but wishing for more detailed coverage are referred to Bollen (1989), a thorough, lucid introduction that treats many topics in SEM with great detail, and to Mueller (1996), an introduction to the topic with emphasis on LISREL and EQS applications. We then present mean structure models, and discuss group comparisons with structural equation models. Finally, we examine with an example how mean and covariance comparison is done in SEM.

8.2 STATISTICAL BACKGROUND

Research applying SEM involves five steps:

1. specification;
2. identification;
3. estimation;
4. modification;
5. interpretation.

That is, a data analyst must first specify a structural equation model representing the research questions, make certain that this model is *identified* in the sense that all parameters have a unique solution, choose a method to estimate the model, modify the model in light of the results if necessary, and interpret the results if the model provides a satisfactory fit to the data. Often step 4 leads back to step 1. For this reason, fitting a structural equation model often can be an iterative process. We brief discuss each of the five steps below.

8.2.1 Notation and Specification

There are two types of variables in SEM, latent and observed. Latent constructs can only be measured through observed variables. The reason we need SEM is the presence of multiple equations in a model. Within a structural equation model, all the equations with latent variables as independent variables and observed variables as dependent variables are collectively known as the measurement model, and all the equations with a latent variable as the dependent variable and at least one independent variable, latent or observed, are collectively known as the latent variable model (or the conceptual or construct model). Sometimes not all concepts are latent—some are directly measured by a single indicator. Thus, the equations describing relations among concepts, be they represented by latent or by observed variables, are known collectively as the structural model.

The latent variables are divided into two groups, depending on their relations to the equation system. If a latent variable is not a dependent

STATISTICAL BACKGROUND

variable anywhere in the system (i.e., is predetermined), it is *exogenous*, and is represented by ξ. If, on the other hand, a latent variable is a dependent variable in at least one of the equations in the system, then it is *endogenous*, and is represented by η. For example, we may be interested in how socioeconomic status (SES) affects cognitive ability. SES, ξ_1, is an exogenous latent variable indicated by four observed variables x_1 to x_4, and cognitive ability, η_1, is an endogenous latent variable indicated by two aptitude test scores y_1 and y_2. Thus, there is only one equation in the latent variable model, that is,

$$\eta_1 = \gamma_{11} \xi_1 + \zeta_1,$$

where γ_{11} is the regression coefficient (the first subscript refers to the sequence number of the dependent variable, and the second refers to the sequence number of the independent variable), and ζ_1 is a random error for the endogenous variable η_1. Suppose that we have another latent variable, academic achievement, which is determined by SES and cognitive ability; then we have another equation in the latent variable model:

$$\eta_2 = \beta_{21} \eta_1 + \gamma_{21} \xi_1 + \zeta_2,$$

where β_{21} is the regression coefficient of η_2 regressed on η_1. There are six equations in the measurement model of the two latent variables η_1 and ξ_1 (let us ignore η_2 for now):

$$x_1 = \lambda_{11} \xi_1 + \delta_1,$$
$$x_2 = \lambda_{21} \xi_1 + \delta_2,$$
$$x_3 = \lambda_{31} \xi_1 + \delta_3,$$
$$x_4 = \lambda_{41} \xi_1 + \delta_4,$$
$$y_1 = \lambda_{11} \eta_1 + \epsilon_1,$$
$$y_2 = \lambda_{21} \eta_1 + \epsilon_2,$$

where the λ's are coefficients or factor loadings for the latent variables, and the δ's and ϵ's are random errors for the observed exogenous and endogenous variables, respectively. There are no intercepts in the equations, because all observed variables are assumed to be measured in deviation form (i.e., deviations from their respective means). The elements of these equations can be collected in vectors or matrices (containing one column per latent variable for the λ's). For the measurement model involving ξ_1 we have

$$\begin{bmatrix} x_1 \\ x_2 \\ x_3 \\ x_4 \end{bmatrix} = \begin{bmatrix} \lambda_{11} \\ \lambda_{21} \\ \lambda_{31} \\ \lambda_{41} \end{bmatrix} [\xi_1] + \begin{bmatrix} \delta_1 \\ \delta_2 \\ \delta_3 \\ \delta_4 \end{bmatrix}$$

where the matrix Λ_x has only one column and the vector ξ has only one element because we have only one ξ-variable. The measurement equation can be expressed in general matrix algebra notation as

$$\mathbf{X} = \Lambda_x \xi + \delta, \tag{8.1}$$

where Λ_x contains parameters for the x's specifically. Similarly, we collect the terms related to the y's in matrix form as

$$\begin{bmatrix} y_1 \\ y_2 \end{bmatrix} = \begin{bmatrix} \lambda_{11} \\ \lambda_{21} \end{bmatrix} [\eta] + \begin{bmatrix} \epsilon_1 \\ \epsilon_2 \end{bmatrix},$$

and more generally in matrix algebra notation as

$$\mathbf{Y} = \Lambda_y \eta + \epsilon. \tag{8.2}$$

In an identical manner, we express the structural equation conveniently as

$$\eta = \mathbf{B}\eta + \Gamma\xi + \zeta, \tag{8.3}$$

and the equation can be reexpressed as

$$\begin{bmatrix} \eta_1 \\ \eta_2 \end{bmatrix} = \begin{bmatrix} 0 & 0 \\ \beta_{21} & 0 \end{bmatrix} \begin{bmatrix} \eta_1 \\ \eta_2 \end{bmatrix} + \begin{bmatrix} \gamma_{11} \\ \gamma_{21} \end{bmatrix} [\xi_1] + \begin{bmatrix} \zeta_1 \\ \zeta_2 \end{bmatrix},$$

which combines the two equations involving η_1 and η_2 given earlier.

There are four variance–covariance matrices: Θ_δ of measurement error terms associated with the observed exogenous variables, Θ_ϵ of measurement error terms associated with the observed endogenous variables, Φ of the latent exogenous variables, and Ψ of error terms of the latent endogenous variables. For the current example,

$$\Phi = \begin{bmatrix} \sigma_{\xi_1}^2 \end{bmatrix},$$

$$\Psi = \begin{bmatrix} \sigma_{\zeta_1}^2 & \\ 0 & \sigma_{\zeta_2}^2 \end{bmatrix},$$

$$\Theta_\delta = \begin{bmatrix} \sigma_{\delta_1}^2 & & & \\ 0 & \sigma_{\delta_2}^2 & & \\ 0 & 0 & \sigma_{\delta_3}^2 & \\ 0 & 0 & 0 & \sigma_{\delta_4}^2 \end{bmatrix},$$

$$\Theta_\epsilon = \begin{bmatrix} \sigma_{\epsilon_1}^2 & \\ 0 & \sigma_{\epsilon_2}^2 \end{bmatrix}.$$

Note that there is only one element in Φ, because there is only one latent exogenous variable, and all variance–covariance matrices are symmetric. That means only the elements in the lower diagonal must be written out. For now the covariance between all the error terms for either the latent or the observed variables are assumed to be zero.

Let us use NX to indicate the number of observed exogenous variables; NY, the number of observed endogenous variables; NK, the number of latent exogenous variables; and NE, the number of latent endogenous variables. The dimensions of the four variance–covariance matrices are $NK \times NK$ for Φ, $NE \times NE$ for Ψ, $NX \times NX$ for Θ_δ, and $NY \times NY$ for Θ_ϵ. The dimensions of the four parameter matrices defined earlier can also be expressed similarly. They are $NX \times NK$ for Λ_x, $NY \times NE$ for Λ_y, $NE \times NE$ for $\mathbf{B}$, and $NE \times NK$ for Γ. In sum, in SEM the patterns of 0 and nonzero elements of the eight matrices of $\mathbf{B}$, Γ, Λ_x, Λ_y, Φ, Ψ, Θ_δ, and Θ_ϵ completely specify a model. For example, for a classical path model where there are no latent variables, $\Lambda_x = \mathbf{I}$, $\Lambda_y = \mathbf{I}$, $\Theta_\delta = \mathbf{0}$, and $\Theta_\epsilon = \mathbf{0}$. Similarly, when $\mathbf{B} = \mathbf{0}$, $\Gamma = \mathbf{0}$, $\Psi = \mathbf{0}$, $\Lambda_y = \mathbf{0}$, and $\Theta_\epsilon = \mathbf{0}$, the general model reduces to a confirmatory factor analysis.

Much of SEM is based on the hypothesis

$$\Sigma = \Sigma(\theta). \tag{8.4}$$

where Σ is the population covariance matrix of observed variables, θ is the vector of model parameters, and $\Sigma(\theta)$ is the model-implied covariance matrix, or the covariance matrix written as a function of θ. Put differently, the estimation of SEM aims to obtain parameters that minimizes the difference between the model-implied variance–covariance matrix $\Sigma(\theta)$ and the observed variance–covariance matrix $\mathbf{S}$.

8.2.2 Identification

Without resorting to mathematical proofs of the matrix $\Sigma(\theta)$ as a function of observed variances and covariances, there are some general rules we can follow. In this section we only deal with the identification of recursive models (models with asymmetric relations other than correlations among exogenous variables). For identification of nonrecursive models refer to Bollen (1989). For recursive models, it suffices to ensure the following:

1. The total number of nonredundant variances and covariances among the $NX + NY$ observed variables is greater than the number of free parameters K in the eight matrices of $\mathbf{B}$, Γ, Λ_x, Λ_y, Φ, Ψ, Θ_δ, and Θ_ϵ.
2. All latent variables must have an assigned unit of measurement. This can be achieved in one of the two ways: A latent variable is scaled to one of its indicator variables, or it is standardized to have unit variance and value 1.

3. If a latent variable has only one indicator, then the latent variable is assumed to be measured without error.
4. The researcher must make use of any relevant rules of identification, such as the two-step rule, the multiple-indicator multiple-cause (MIMIC) rule, and Wald's rank rule (Bollen 1989) to further ensure identification of the model.

8.2.3 Estimation

The variance–covariance matrix Σ of the observed variables in general SEM is

$$\Sigma = E\left(\begin{bmatrix} Y \\ X \end{bmatrix}\begin{bmatrix} Y \\ X \end{bmatrix}'\right) = E\left(\begin{bmatrix} YY' & YX' \\ XY' & XX' \end{bmatrix}\right), \quad (8.5)$$

where

$$\begin{bmatrix} Y \\ X \end{bmatrix}' = [Y_1 \; Y_2 \; \cdots \; Y_{NY} \; X_1 \; X_2 \; \cdots \; X_{NX}].$$

Replacing the elements of the last matrix of (8.5) with latent variables and associated parameters and error terms, we can show that the model-implied variance–covariance matrix $\Sigma(\theta)$ for a general structural equation model can be expressed in terms of the eight matrices as (Mueller 1996, p. 152)

$$\Sigma(\theta) = \begin{bmatrix} \Lambda_y(I-B)^{-1}(\Gamma\Phi\Gamma' + \Psi)\left[(I-B)^{-1}\right]'\Lambda_y' + \Theta_\epsilon & \Lambda_y(I-B)^{-1}\Gamma\Phi\Lambda_x' \\ \Lambda_x\Phi\Gamma'\left[(I-B)^{-1}\right]'\Lambda_y & \Lambda_x\Phi\Lambda_x' + \Theta_\delta \end{bmatrix}. \quad (8.6)$$

From (8.6) we may derive simplified expressions for the path model or the confirmatory factor model where some of the eight matrices are identity or null matrices.

The aim of estimation of an identified structural equation model is to obtain estimates $\hat{B}$, $\hat{\Gamma}$, $\hat{\Lambda}_x$, $\hat{\Lambda}_y$, $\hat{\Phi}$, $\hat{\Psi}$, $\hat{\Theta}_\delta$, and $\hat{\Theta}_\epsilon$ that collectively contribute to the model-implied variance–covariance matrix $\Sigma(\hat{\theta})$ so that $\Sigma(\hat{\theta})$ approximates Σ. This is achieved iteratively by gradually finding estimates that provide better and better approximation. To measure this approximation, a fitting function in the form of $F[S\Sigma(\theta)]$ is evaluated at every iteration to ensure that the distance between S (the sample estimate of Σ) and $\Sigma(\hat{\theta})$ decreases until a minimum is found when the value of the fitting function changes less from the previous iteration than a predetermined criterion.

Two of the most common methods of estimation are ML and generalized least squares (GLS). Both ML and GLS rely on the assumption of the multivariate normal distribution of the data. The assumption may be too

STATISTICAL BACKGROUND   117

restrictive for SEM analysis at times. To deal with the problem, SEM software packages such as LISREL and EQS implement a flexible technique of estimation known as the asymptotically distribution free (ADF) method, based on a fourth-order multivariate product moment of the variables (Browne 1984).

Between the more restrictive estimation methods such as ML and GLS (requiring the assumption of the multivariate normal distribution) and ADF type methods (requiring much less) lies another estimation method based on elliptical distribution theory. The multivariate elliptical is a class of distributions, of which the multivariate normal is a special case. We should note that both ML and GLS estimation methods have their counterparts in the multivariate elliptical form as well as in the ADF form. The choice gives the analyst more flexibility in choosing an appropriate estimation method.

8.2.4 Modification

An initially hypothesized model has been specified, identified, and estimated. Assuming the model estimation has converged and that there is no indication of errors in the estimation, the data analyst is now faced with sheets (or screens) of output. The best outcome is a perfectly confirmed hypothesis in that the initially hypothesized model fits the data very well. Oftentimes, however, the outcome is not that simple. The hypothesized model may fit the data poorly, judging by any of the plethora of fitting indices available today. The researcher now should go back to the drawing-board and start at step 1 again.

In rethinking the initial model, there are at least four aspects of the model that can be improved:

1. First, is the initial model too complex in the sense that certain parameters may not be necessary? Researchers often make the model too complicated even as a first try. Some parameters may not be justified either statistically or substantively.
2. Are there paths or parameters that one should be free to estimate but that were fixed in the initial model? Software such as LISREL provides modification indices to facilitate the assessment of potentially worthy parameters that are currently fixed.
3. Is the initial model correctly specified, not in simple terms of parameters and paths, but in additional variables, both observed and latent?
4. Coming from a linear regression tradition, a common tendency for first-time SEM analysts is to include variables representing fixed groups such as race, gender, time period, or age cohort as one of the many exogenous variables. Can some of these be considered as grouping variables? In other words, is a multiple group or population analysis more appropriate?

Answers to these questions will guide us to a more sensible specification of models in the next effort to analyze the data. The last point in particular, as the focal point in this chapter, is discussed more thoroughly later on.

8.2.5 Interpretation

Once a satisfactory model is arrived at, the final task is to make sense of the results. While the ways of interpreting both standardized and unstandardized estimates are similar to those in linear regression as far as direct effects are concerned, the complexity of a structural equation model presents further challenges. For example, in addition to the direct effect of SES on academic achievement, SES also has an indirect effect through cognitive ability. The sum of its direct effect and indirect effect, then, form its total effect on academic achievement.

To formalize, the matrix $\mathbf{B}$ contains all direct effects among the η-variables in a model. Their indirect effects (IE) are

$$\text{IE}_{\eta\eta} = \mathbf{B}^2 + \mathbf{B}^3 + \mathbf{B}^4 + \cdots, \tag{8.7}$$

where $\mathbf{B}^2$ is the matrix product $\mathbf{BB}$, and so on. The maximal number of intervening variables between any pair of η's determines how high the powers in (8.7) should go. Put differently, the maximal number of paths between any pair of η's equals the highest power in (8.7).

Since the total effect (TE) equals the sum of the direct and indirect effects, it follows that

$$\text{TE}_{\eta\eta} = \mathbf{B} + \mathbf{B}^2 + \mathbf{B}^3 + \mathbf{B}^4 + \cdots. \tag{8.8}$$

Similarly, we may obtain the direct, indirect, and total effects between latent exogenous and endogenous variables. Because direct effects between them are contained in the matrix $\mathbf{\Gamma}$, we must multiply $\mathbf{\Gamma}$ and $\mathbf{B}$ to get IE:

$$\text{IE}_{\eta\xi} = (\mathbf{B} + \mathbf{B}^2 + \mathbf{B}^3 + \mathbf{B}^4 + \cdots)\mathbf{\Gamma}. \tag{8.9}$$

To obtain their total effects, we must include $\mathbf{\Gamma}$ one more time to account for direct effects, thereby obtaining

$$\text{TE}_{\eta\xi} = (\mathbf{I} + \mathbf{B} + \mathbf{B}^2 + \mathbf{B}^3 + \mathbf{B}^4 + \cdots)\mathbf{\Gamma}. \tag{8.10}$$

To make a sensible interpretation, just as in classical path analysis, the structural effect components presented above work with standardized parameters only.

8.3 MEAN AND COVARIANCE STRUCTURES

So far we have considered covariance structures only. That is, we have ignored latent-variable means and measurement-equation intercepts. In the typical SEM application, the parameters of a structural equation are the regression coefficients, and the covariance structure of the variables bears the most important parametric information. In that way, the means of the observed variables are not decomposed into parameters to estimate. This is fine when the researcher is interested in slope coefficients only, either for assessing the effects of certain (latent) structural variables or for comparing slope effects across groups. In such cases the model is specified, estimated, and tested via the sample covariance matrix $\mathbf{S}$.

However, there are situations where the data analyst wishes to know the means and intercepts. When conducting group comparisons in particular, a researcher may be interested in comparing the means of the same latent variables in different groups. When this occurs, the sample mean vector $[\bar{\mathbf{Y}}\ \bar{\mathbf{X}}]'$ too carries important statistical information, and both the sample covariance matrix $\mathbf{S}$ and the sample mean vector $[\bar{\mathbf{Y}}\ \bar{\mathbf{X}}]'$ must be analyzed simultaneously.

To help understand the relation between the means of variables and the intercept in an equation, let us consider a simple bivariate regression model. In notation consistent with Chapter 3, a linear regression with one explanatory variable takes the form

$$y = \beta_0 + \beta_1 x + e,$$

where β_0 is the intercept. The intercept helps define the mean of y, though it is not equal to the mean in general. As usual, the mean of the error term is assumed to be zero. Then by taking expectations of both sides we get

$$\mu_y = \beta_0^* + \beta_1 \mu_x,$$

where μ_y is the mean of y, μ_x is the mean of x, and β_0^* is the new intercept, which is the value of y when x is at its mean but is identical to β_0 if the variables are in deviation form with zero as the origin. In this expression, μ_y is given in terms of β_0^*, β_1, and μ_x. The expression illustrates the approach known as the *structured means* model, incorporating a *mean structure* into the covariance analysis.

Recall that all random variables were assumed to have zero means in the three principal SEM equations (8.1), (8.2), and (8.3). Following the discussion above, we now relax this assumption and extend (8.1), (8.2), and (8.3) to include, in addition to the previous eight, four new parameter matrices containing intercepts and means of the latent variables. Where there is a discrepancy in notation in the SEM literature, the LISREL notational convention is followed here.

Corresponding to (8.1), (8.2), and (8.3), the three basic equations accommodating an intercept matrix become

$$X = \tau_x + \Lambda_x \xi + \delta, \tag{8.11}$$

$$Y = \tau_y + \Lambda_y \eta + \epsilon, \tag{8.12}$$

$$\eta = \alpha + B\eta + \Gamma\xi + \zeta, \tag{8.13}$$

where τ_x, τ_y, and α are conceptually equivalent to the β_0 in the simple bivariate regression intercept, though they are now vectors containing intercepts with dimensions equal to Y ($NY \times 1$), X ($NX \times 1$), and η ($NE \times 1$), respectively. Because we assume, as before, that the expected values of the error terms are zero, we may further obtain the means of the latent variables. We do need, however, a final matrix, κ, for the mean of ξ, or $E(\xi)$. Thus, by taking the expectation of (8.13) we find the mean of η, $E(\eta)$, as

$$E(\eta) = (I - B)^{-1}(\alpha + \Gamma\kappa), \tag{8.14}$$

where κ has the same dimension as ξ ($NK \times 1$). Similarly, we define the means of X and Y by

$$\mu_x = \tau_x + \Lambda_x \kappa, \tag{8.15}$$

$$\mu_y = \tau_y \Lambda_y (I - B)^{-1}(\alpha + \Gamma\kappa). \tag{8.16}$$

In LISREL, the four additional matrices of τ_x, τ_y, α, and κ are used explicitly in the program and are assumed by default as fixed zero matrices. This means that by default LISREL estimates a covariance structure model unless some of the four intercept–mean matrices are set differently. By setting some (or all) of the four matrices free, a mean and covariance structure model can be estimated. In EQS, mean structures are handled by including a constant "variable" V999 and requesting an analysis of the moment matrix. The minimum data requirement includes, in addition to the covariance (or correlation plus standard deviation), vectors of means for the observed variables.

In general, in a single sample analysis, all the mean parameters of τ_x, τ_y, α, and κ will not be identified without further conditions imposed. In principle, the number of intercepts to be estimated should be less than or equal to the number of observed variables. In the analysis of multiple groups (samples or populations), more means can be estimated than in a single-group analysis, as illustrated in a later section.

Recall that for estimating a covariance structure model, one basic condition is that all latent variables have their scales (or measurement units) set, say, to one of their respective indicators. In an identical sense, latent-variable means are not identified until their origins are specified. Let us use the SES example to illustrate. For measuring SES, we have four indicators given earlier in the chapter. Applying (8.15) and writing it out in scalar form, we have

$$E(\dot{x}_1) = 0 + 1\kappa_1,$$
$$E(x_2) = \tau_{x_2} + \lambda_{21}\kappa_1,$$
$$E(x_3) = \tau_{x_3} + \lambda_{31}\kappa_1,$$
$$E(x_4) = \tau_{x_4} + \lambda_{41}\kappa_1.$$

Here the scale of ξ_1 (SES) is set to that of x_1 (e.g., mother's education), thereby rendering the other λ's identified. Furthermore, the zero point of ξ_1 is set the same as that of x_1, thus making κ_1 equal to $E(x_1)$. Consequently, all other τ's are identified. The choices are, of course, arbitrary. We could set another λ to 1 and another τ_x to 0 instead. Once the mean (κ_1) of the latent exogenous variable SES (ξ_1) is identified, $\boldsymbol{\alpha}$ and then the mean of $\boldsymbol{\eta}$ will become identified.

8.4 GROUP COMPARISON IN SEM

As with other statistical methods, multiple group comparison in SEM can be quite useful, whether the groups are fixed, chosen at random (as in experimental designs) or assigned nonrandomly (as in quasiexperimental designs). However, because of the complexity of SEM in terms of the potential total number of variables, both observed and latent, and in terms of the relationships between the observed and latent variables, and the exogenous and endogenous variables, with or without the complication of correlated error terms, there are many ways in which to compare the groups.

First of all, equality constraints among groups can be tested for the eight matrices $\boldsymbol{\Lambda}_x$, $\boldsymbol{\Lambda}_y$, $\mathbf{B}$, $\boldsymbol{\Gamma}$, $\boldsymbol{\Phi}$, $\boldsymbol{\Psi}$, $\boldsymbol{\Theta}_\delta$, and $\boldsymbol{\Theta}_\epsilon$ in covariance structure models. Furthermore, hypotheses of across-group parameter equality of the four matrices $\boldsymbol{\tau}_x$, $\boldsymbol{\tau}_y$, $\boldsymbol{\alpha}$, and $\boldsymbol{\kappa}$ for mean structures can be tested. Thus, we may form hypotheses comparing groups for some or all of the twelve parameter matrices

$$\boldsymbol{\Lambda}_{gx}, \boldsymbol{\Lambda}_{gy}, \mathbf{B}_g, \boldsymbol{\Gamma}_g, \boldsymbol{\Phi}_g, \boldsymbol{\Psi}_g, \boldsymbol{\Theta}_{g\delta}, \boldsymbol{\Theta}_{g\epsilon}, \boldsymbol{\tau}_{gx}, \boldsymbol{\tau}_{gy}, \boldsymbol{\alpha}_g, \boldsymbol{\kappa}_g,$$

where the subscript g indicates the gth group for $g = 1, 2, \ldots, G$. To be consistent with the notation indicating groups in this book, the subscript g is used, though in the SEM literature both it and superscript (g) are found. If a researcher is interested in testing a hypothesis of parameter equality between two groups for all the structural relationships, similar to those in the GLM, for example, the null hypothesis will be

$$H_0: \quad \mathbf{B}_g = \mathbf{B}_h \wedge \mathbf{\Gamma}_g = \mathbf{\Gamma}_h$$

where $g \neq h$ and $\wedge$ means "and" and implies a joint test.

However, the many aspects of SEM in its distribution assumptions and model specification suggest that tests of group invariance should also be made in them, not just in parameters. These aspects are multivariate distributions of the observed variables, the covariance matrices $\mathbf{\Sigma}$, the form of the structural equation model, and the model parameters $\mathbf{\theta}$. We examine these four topics below.

8.4.1 Equality of Multivariate Distributions

The most common distribution assumption underlying SEM is the multivariate normal distribution. Despite its restricted nature and its likely violation by many sources of data, especially in the social sciences, the multivariate normal distribution still is assumed in the majority of SEM applications because it is a simple extension of the normal distribution, which underlies the classical linear regression model. Recent advances in SEM methodology have made possible alternative distribution assumptions, as previously discussed in the section on estimation. The multivariate elliptical distribution and the ADF method can be viewed as distribution assumptions that are less restrictive. These advances also have made it possible to question the assumption of equal distribution across groups. That is, would the data from one group satisfy the multivariate normal distribution while the data from another would not?

To answer a question like this, a data analyst may cast a hypothesis about the multivariate distribution of the observed data. Let $\mathbf{Z} = [\mathbf{X}\ \mathbf{Y}]$, and md($\mathbf{Z}$) denote the multivariate distribution of $\mathbf{Z}$. One may be interested in testing

$$H_0: \quad \text{md}(Z_1) = \text{md}(Z_2) = \cdots = \text{md}(Z_G).$$

That is, the null hypothesis states that all groups have multivariate distributions.

There is no formal test for this hypothesis. A reasonable way to test it less formally is to fit a certain structural equation model of interest to the data from all the groups separately. By keeping the model specification identical for all the groups while fitting the model using estimation methods based on the multivariate normal distribution, the multivariate elliptical distribution, and the ADF, a researcher may compare across groups certain model fitting indices such as the model χ^2 or BIC from the same group under two or more of these distribution assumptions. A larger difference between the model fitting indices from different distribution assumptions for the same group than the same difference in another group indicates potential difference in the underlying multivariate distribution. How can we conclude that the difference between two groups is large enough to suggest different multivariate distributions? One possible tool for making the decision is to transform the model χ^2 into the BIC, and then the BIC into the Bayes factor. The guideline for comparing models based on posterior odds presented in Table 7.8 should serve for determining if the discrepancy between two distribution assumption from group g is greater than that from group h, thereby suggesting different underlying distributions.

8.4.2 Equality of Covariance Matrices

When comparing across groups in SEM, another difference can be that of the covariance matrix across the groups of interest. Thus, one possible hypothesis is that all covariance matrices are equal, namely,

$$H_0: \quad \Sigma_1 = \Sigma_2 = \cdots = \Sigma_G.$$

The hypothesis is not very useful, because we only observe $\mathbf{S}_g$, and differences in the sample covariance matrices can be due to the difference in sample variances. Here what we are really interested in is the equality among the correlation matrices. The covariance matrices may be unequal because the sample variances may not be the same for the variables in all G groups. As long as the correlation matrices are equal, the basis for structural relations will remain equal.

Let us denote the population correlation matrix for group g as $\mathbf{C}_g$, and relate it to Σ_g by

$$\Sigma_g = \sigma'_g \mathbf{C}_g \sigma_g$$

where σ_g is the diagonal matrix of standard deviations of the variables in $\mathbf{Z}$. The equality hypothesis to test then is

$$H_0: \quad \mathbf{C}_1 = \mathbf{C}_2 = \cdots = \mathbf{C}_G.$$

To test the hypothesis, one can consider all X and Y variables as X-variables, and analyze the input covariance matrix. For each x-variable, specify a ξ without the error term in the measurement equation

$$\mathbf{X} = \boldsymbol{\Lambda}_x \boldsymbol{\xi},$$

where the estimated parameter matrix $\boldsymbol{\Lambda}_x$ equals $\boldsymbol{\sigma}$; that is, the nonzero diagonal elements are just the standard deviations of the variables. All that remains to do is a test of equality among the correlation matrices of ξ_g

$$H_0: \quad \boldsymbol{\Phi}_1 = \boldsymbol{\Phi}_2 = \cdots = \boldsymbol{\Phi}_G.$$

By testing this hypothesis we are testing the equality of the correlation matrices while allowing the sample variances to be different. If, however, a data analyst believes that the variances should be equal among the groups, the Λ_{gx} (really the standard deviations) can be constrained to be equal. One then tests a more restrictive hypothesis that all the population Σ_g are equal across groups.

8.4.3 Equality of Model Forms

Certain differences in model forms may be specified by constraining parameters to be zero, thus removing paths or even variables from a model. Such differences in model forms could be considered as due to equality of model parameters, as discussed in the next subsection. Sometimes, however, a difference in model form cannot be easily specified as a constraint on parameters. Thus, following Bollen (1989), we define two models to have the same *form* if both have the same parameter matrices with the same dimensions and the same location of fixed, free, and constrained parameters. By using this definition, we can confine the topic of equality of model parameters to the situation where only values of free parameters can compared. In most applications researchers assume that the form of two models is identical, and they concentrate on the similarity of parameter values within a given form.

To illustrate different forms for two groups, let us suppose that the concept of interest is one's intensity of social interaction. x_1, x_2, x_3, and x_4 represent interaction with immediate family, other relatives, friends, and neighbors respectively. For one population, the model to be specified may be

$$\begin{bmatrix} x_1 \\ x_2 \\ x_3 \\ x_4 \end{bmatrix} = \begin{bmatrix} 1 \\ \lambda_{21} \\ \lambda_{31} \\ \lambda_{41} \end{bmatrix} [\xi_1] + \begin{bmatrix} \delta_1 \\ \delta_2 \\ \delta_3 \\ \delta_4 \end{bmatrix}$$

while for another population the model may contain a second latent variable to separate kin social interaction from nonkin social interaction, thus increasing the dimension of the matrices Λ_x and ξ:

$$\begin{bmatrix} x_1 \\ x_2 \\ x_3 \\ x_4 \end{bmatrix} = \begin{bmatrix} 1 & 0 \\ \lambda_{21} & 0 \\ 0 & 1 \\ 0 & \lambda_{41} \end{bmatrix} \begin{bmatrix} \xi_1 \\ \xi_2 \end{bmatrix} + \begin{bmatrix} \delta_1 \\ \delta_2 \\ \delta_3 \\ \delta_4 \end{bmatrix}$$

where ξ_1 and ξ_2 are correlated with the parameter ϕ_{21}. The possibility of having different forms may provide better fit to the data from different populations.

Another possibility of difference in model form includes the level of latent variables. For example, it may be that in one model only first-order latent variables are present while in another model second-order or even higher-order latent variables may be present. By using the intensity of social interaction as a latent endogenous variable in a general structural equation model (and by changing the x's to y's and the ξ's to η's etc.), the model for the first group has only one first-order latent variable representing social interaction, while the model for the second group may contain a second-order latent variable of general social interactions that are represented by two specific sources of social interaction with kin and nonkin relations.

Testing equality of model forms should be straightforward. When models are of different forms, they are often imbedded within one another. The LRT and its asymptotic equivalents should then apply. However, sometimes the models to compare are not imbedded within one another. Bayesian model comparison methods can then be used, such as posterior odds based on Bayes factors.

8.4.4 Equality of Model Parameters

Assuming models are of the same form, we may further test to see if the model parameters are equal. The types of equality hypothesis to consider fall into three categories: hypotheses about the structural (conceptual) model, hypotheses about the measurement model, and hypotheses about the general structural equation model. Following Bollen's (1989) discussion of increasing restriction in testing equality, we present three hierarchies below.

8.4.4.1 Testing Equal Structural Models

Oftentimes the researcher is interested in the structural parameters only. Assuming the model forms in both the structural and the measurement models are identical across groups, the natural next step is to see whether the structural parameters are equal across groups; if they are, whether the covariance matrices of the structural error terms are equal; and finally, whether the correlations of latent exogenous variables are also equal:

$$H_{\mathbf{B}\Gamma}: \quad \mathbf{B}_g = \mathbf{B}_h \wedge \mathbf{\Gamma}_g = \mathbf{\Gamma}_h,$$

$$H_{\mathbf{B}\Gamma\Psi}: \quad \mathbf{B}_g = \mathbf{B}_h \wedge \mathbf{\Gamma}_g = \mathbf{\Gamma}_h \wedge \mathbf{\Psi}_g = \mathbf{\Psi}_h,$$

$$H_{\mathbf{B}\Gamma\Psi\Phi}: \quad \mathbf{B}_g = \mathbf{B}_h \wedge \mathbf{\Gamma}_g = \mathbf{\Gamma}_h \wedge \mathbf{\Psi}_g = \mathbf{\Psi}_h \wedge \mathbf{\Phi}_g = \mathbf{\Phi}_h,$$

where g and h represent different groups. Each hypothesis can be evaluated by some model fitting statistics.

8.4.4.2 Testing Equal Measurement Models

Sometimes the researcher may be interested in the measurement parameters as well. One can move on to test and see whether the measurement parameters are equal across groups; if they are, whether the covariance matrices of the measurement error terms are equal; and finally, whether the measurement parameters, errors, and correlations of the latent exogenous variables are jointly equal between groups:

$$H_{\Lambda_x}: \quad \Lambda_{gx} = \Lambda_{hx},$$

$$H_{\Lambda_x \Theta_\delta}: \quad \Lambda_{gx} = \Lambda_{hx} \wedge \Theta_{g\delta} = \Theta_{h\delta},$$

$$H_{\Lambda_x \Theta_\delta \Phi}: \quad \Lambda_{gx} = \Lambda_{hx} \wedge \Theta_{g\delta} = \Theta_{h\delta} \wedge \Phi_g = \Phi_h.$$

In certain applications of general SEM, the researcher may be interested in equal measurement parameters only, instead of equal structural effects. The hypotheses given above should apply there too.

8.4.4.3 Testing Equal General Models

Finally, in general SEM, one may be interested in both the measurement model and the structural model, thus wanting to test for equality in both. One may, for example, be interested in testing the equality hypotheses

$$H_{\mathrm{B}\Gamma}, H_\Lambda, H_{\Lambda\mathrm{B}\Gamma}, H_{\Lambda\mathrm{B}\Gamma\Psi}, H_{\Lambda\mathrm{B}\Gamma\Theta}, H_{\Lambda\mathrm{B}\Gamma\Psi\Theta}, H_{\Lambda\mathrm{B}\Gamma\Psi\Theta\Phi},$$

where Λ represents both Λ_x and Λ_y, and Θ represents both Θ_δ and Θ_ϵ.

Because of the frequent need for testing equalities like these, all major SEM software has built-in function for testing equality constraints for model parameters across groups, often using LMT. This facility is illustrated with EQS and LISREL in the example section.

8.4.4.4 Testing Equal Mean Structures

The discussion above focuses on testing equality among covariance structures. It is possible to test equality among mean structures. This simply involves setting up hypotheses specific for the four intercept–mean matrices. We may have

$$H_{\tau_x \tau_y}: \quad \tau_{gx} = \tau_{hx} \wedge \tau_{gy} = \tau_{hy},$$

$$H_{\alpha\kappa}: \quad \alpha_g = \alpha_h \wedge \kappa_g = \kappa_h,$$

$$H_{\tau_x \tau_y \alpha \kappa}: \quad \tau_{gx} = \tau_{hx} \wedge \tau_{gy} = \tau_{hy} \wedge \alpha_g = \alpha_h \wedge \kappa_g = \kappa_h.$$

One may combine these to test customized hypotheses. The mean latent scores can be of particular interest to certain researchers. Earlier we discussed the need to scale the mean of a latent variable with an observed intercept. When making group comparisons, however, this may not be necessary if we use the means of the latent variables in one group as the benchmark (i.e., origin) for comparison, as is illustrated in the example section below. Hypotheses of equal mean structures can be combined with those of equal covariance structures in many different ways, as the researcher desires.

8.5 AN EXAMPLE

In this section we examine a two-group example applying SEM. In the U.S., Head Start is a well-known educational program for children who might be academically behind. Sörbom (1981) reanalyzed a sample of 303 white children from the Head Start summer program, which consists of a Head Start sample ($N_1 = 148$) and a matched control sample ($N_2 = 155$). In this section we compare the two samples in terms of their correlation matrices, distribution assumptions for estimation, covariance structures, and mean structures. The comparison of their mean structures will allow us to assess the mean latent scores for cognitive ability so that we can determine if the program was successful. However, we must control for the children's SES backgrounds. Because both cognitive ability and SES are measured with multiple indicators, SEM is called for.

Just as presented earlier in the chapter, SES is measured by four observed variables and cognitive ability by two:

x_{g1}: Mother's education (MaEd)
x_{g2}: Father's education (PaEd)
x_{g3}: Father's occupation (PaOcc)
x_{g4}: Family income (FamInc)
y_{g1}: Score on the Metropolitan Readiness Test (MRT)
y_{g2}: Score on the Illinois Test of Psycholinguistic Abilities (ITPA)

The subscript g indicates membership in the Head Start group ($g = 1$) or the control group ($g = 2$). The six indicators are presented in Tables 8.1 and 8.2 for the Head Start group and the control group, respectively. For ease of copying the data into computer programs, we intentionally keep the correlation matrices in two tables instead of as two off-diagonal halves in one table. The two matrices can then be included in the computer programs given in the appendix (Section 8.A). Means and standard deviations of the variables are also presented in the tables.

8.5.1 Comparing Correlation Matrices

Before we embark on our usual SEM tests of parameter equality, we may be interested in testing the hypothesis of equal correlation matrices:

$$H_0: \quad \mathbf{C}_1 = \mathbf{C}_2.$$

Table 8.1 Means, Standard Deviations, and Correlations for the Head Start Group

Variable	MaEd	PaEd	PaOcc	FamInc	MRT	ITPA
MaEd	1.000					
PaEd	0.441	1.000				
PaOcc	0.220	0.203	1.000			
FamInc	0.304	0.182	0.377	1.000		
MRT	0.274	0.265	0.208	0.084	1.000	
ITPA	0.270	0.122	0.251	0.198	0.664	1.000
$\hat{\mu}$	3.520	3.081	2.088	5.358	19.672	9.562
$\hat{\sigma}$	1.332	1.281	1.075	2.648	3.764	2.677

Note: $N_1 = 148$.

Table 8.2 Means, Standard Deviations, and Correlations for the Control Group

Variable	MaEd	PaEd	PaOcc	FamInc	MRT	ITPA
MaEd	1.000					
PaEd	0.484	1.000				
PaOcc	0.224	0.342	1.000			
FamInc	0.268	0.215	0.387	1.000		
MRT	0.230	0.215	0.196	0.115	1.000	
ITPA	0.265	0.297	0.234	0.162	0.635	1.000
$\hat{\mu}$	3.839	3.291	2.600	6.435	20.415	10.070
$\hat{\sigma}$	1.360	1.195	1.193	3.239	3.900	2.719

Note: $N_2 = 155$.

Using the methods described in an earlier section, the hypothesis can be tested with the EQS and LISREL programs given in Sections 8.A.1 and 8.A.2 at the end of this chapter. The test results are given in Table 8.3, with the LMT of individual correlation equalities from the EQS program output and the overall tests from both the EQS and LISREL program outputs. A diagram for the equal correlation structure is presented in Figure 8.1, generated by LISREL 8.12. The estimates and the curves are a bit crowded in the diagram, due to the sheer number of them. Judging by the LMTs, only ϕ_{52} and ϕ_{62} are significantly different between the two groups. Overall, however, the two correlation matrices are indistinguishable from each other, given the ML $\chi^2 = 10.543$ and 15 degrees of freedom ($p = 0.784$). The ERLS χ^2 gives almost identical conclusions.

AN EXAMPLE

Table 8.3 LM χ^2 of Equality between the Correlation Matrices

Variable	MaEd	PaEd	PaOcc	FamInc	MRT
MaEd					
PaEd	0.107				
PaOcc	0.109	1.475			
FamInc	0.090	0.139	0.014		
MRT	0.021	3.906	0.001	0.903	
ITPA	0.285	6.258	0.118	0.615	0.519
χ^2, ML			10.543		
χ^2, ERLS			10.378		
df			15		

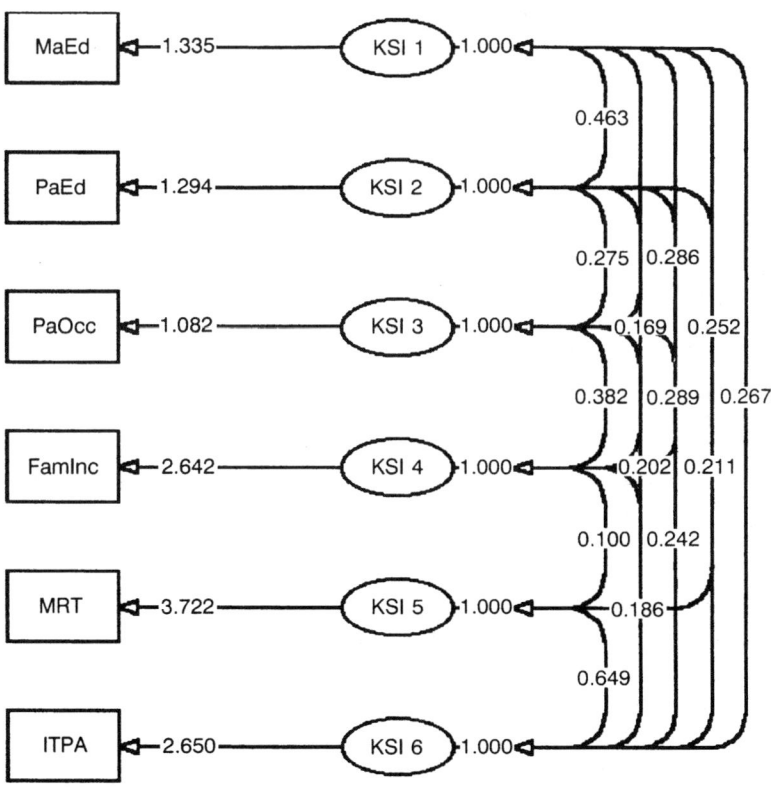

Figure 8.1 Equal correlation matrix for the Head Start and the control groups.

8.5.2 Comparing Covariance Structures and Multivariate Distributions

To comparing the two covariance structures, we must first specify both the measurement and the structure models. We use ξ_{g1} to denote SES and η_{g1} to denote cognitive ability. Thus, the observed variables are related to the latent variables by the following measurement equations:

$$x_{g1} = \lambda_{g11} \xi_1 + \delta_{g1},$$

$$x_{g2} = \lambda_{g21} \xi_1 + \delta_{g2},$$

$$x_{g3} = \lambda_{g31} \xi_1 + \delta_{g3},$$

$$x_{g4} = \lambda_{g41} \xi_1 + \delta_{g4},$$

$$y_{g1} = \lambda_{g11} \eta_1 + \epsilon_{g1},$$

$$y_{g2} = \lambda_{g21} \eta_1 + \epsilon_{g2},$$

where the subscript g is added to indicate group membership. The conceptual model is defined by the structural equation

$$\eta_{g1} = \gamma_{g11} \xi_{g1} + \zeta_{g1},$$

where again the subscript g is added in the structural equation.

Under the model specification above, we estimated three models: the first on the Head Start sample, the second on the control sample, and the third on both samples by constraining most parameters to be equal between the two groups. The results are presented in Table 8.4, and the computer programs for the third model are given in Sections 8.A.3 and 8.A.4. A path diagram of the equal covariance structure model generated by LIREL 8.12 is given in Figure 8.2. Parameter estimates are included for the paths. Note that the errors representing the error terms are not produced.

First, let us examine the multivariate normal assumption underlying the SEM for the two groups separately. By relaxing the multivariate normal assumption to the multivariate elliptical assumption, the χ^2 for the Head Start group improved by about 0.5, and that for the control group improved by about 0.2. The difference between the improvements due to the different distribution assumptions is minute, suggesting the multivariate normality assumption does not present any problem.

From the parameter estimates of the covariance structural analyses for the two groups, some differences can be seen, but there do not seem to be any major discrepancies. To formalize this conjecture, we test the hypothesis

$$H_{\Lambda_1 \Gamma_1 \theta_{\delta_{121}}} = H_{\Lambda_2 \Gamma_2 \theta_{\delta_{221}}},$$

AN EXAMPLE

Table 8.4 ML Estimates from Covariance Structure Models of the Head Start Data

Parameter		Head Start		Control		Equality		
		Estimate	S.e.	Estimate	S.e.	Estimate	S.e.	LM χ^2
Λ_x:	λ_{21}	0.669	0.203	1.033	0.235	0.874	0.152	0.857
	λ_{31}	0.963	0.272	1.259	0.362	1.109	0.224	0.100
	λ_{41}	2.399	0.678	2.793	0.819	2.603	0.530	0.001
Λ_y:	λ_{21}	0.876	0.214	0.885	0.219	0.862	0.150	0.005
Γ:	γ_{21}	1.950	0.729	2.101	0.796	2.049	0.542	0.080
Θ_δ:	θ_{21}	0.469	0.149	0.405	0.140	0.426	0.102	0.127
ML:								
χ^2		15.655		8.756		26.017		
NFI		0.917		0.955		0.932		
NNFI		0.893		0.979		0.974		
ERLS:								
χ^2		15.148		8.572		25.528		
NFI		0.925		0.962		0.940		
NNFI		0.907		0.984		0.979		
df		7		7		20		

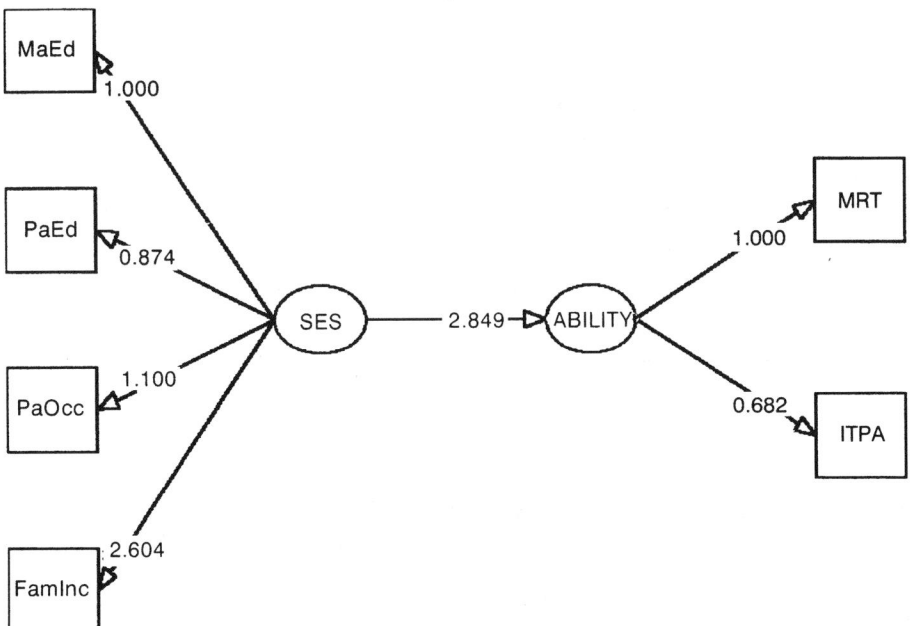

Figure 8.2 Equal covariance structure for the Head Start and the control groups.

where $\theta_{\delta_{121}}$ is singled out for testing because we are not interested in testing equality between the error variances of the x-variables in the measurement equation. Parameters from both the matrices Λ_x and Λ_y and the matrix Γ are assumed to be equal between the two groups. The values of the equal (constrained) parameters are about halfway between the parameter estimates from the Head Start model and those from the control model. Judging by the individual LMT statistics or by the overall LRT ($\chi^2 = 26.017$ with df = 20), there appears to be no difference between the two sets of model parameters.

8.5.3 Comparing Mean and Covariance Structures

Now let us introduce mean structures into the comparison. For doing so, we will need intercepts in both the measurement and the structural equations:

$$x_{g1} = \tau_{g1} + \lambda_{g11}\xi_1 + \delta_{g1},$$
$$x_{g2} = \tau_{g2} + \lambda_{g21}\xi_1 + \delta_{g2},$$
$$x_{g3} = \tau_{g3} + \lambda_{g31}\xi_1 + \delta_{g3},$$
$$x_{g4} = \tau_{g4} + \lambda_{g41}\xi_1 + \delta_{g4},$$
$$y_{g1} = \tau_{g1} + \lambda_{g11}\eta_1 + \epsilon_{g1},$$
$$y_{g2} = \tau_{g2} + \lambda_{g21}\eta_1 + \epsilon_{g2},$$

where τ_{g1} is the intercept for the measurement equation explaining x_{g1} in the gth group. The structural equation is modified accordingly,

$$\eta_{g1} = \alpha_{g1} + \gamma_{g11}\xi_{g1} + \zeta_{g1},$$

where α_{g1} is the intercept for η_{g1}. Notice that κ does not come into the system unless an equation is expressed in terms of the means of the variables, as in (8.14) through (8.16). We again estimated three mean and covariance structure models, one for each group and a third assuming the equality hypothesis

$$H_{\tau_1\Lambda_1\Gamma_1\theta_{\delta_{121}}} = H_{\tau_2\Lambda_2\Gamma_2\theta_{\delta_{221}}},$$

where τ_g is added to reflect the equality subhypothesis for the intercepts. Notice that the matrices α_g and κ_g are not included in the hypothesis for the joint test. To fit an identified model, the latent means for SES and cognitive ability are assumed to be zero for the control group, so that the counterparts in the Head Start group can be scaled with an origin or estimated with a benchmark. The equality of the α's and κ's can be directly assessed from the parameter estimates. The results are presented in Table 8.5, and the computer programs for the third model are given as Sections

Table 8.5 ML Estimates from Mean and Covariance Structure Models of the Head Start Data

Parameter	Head Start Estimate	Head Start S.e.	Control Estimate	Control S.e.	Equality Estimate	Equality S.e.	LM χ^2
τ_x: τ_1	3.520	0.110	3.839	0.110	3.870	0.094	0.033
τ_2	3.081	0.106	3.291	0.096	3.340	0.083	0.772
τ_3	2.088	0.089	2.600	0.096	2.574	0.090	0.582
τ_4	5.358	0.218	6.435	0.261	6.421	0.228	0.011
Λ_x: λ_{21}	0.669	0.203	1.033	0.235	0.853	0.144	1.455
λ_{31}	0.963	0.272	1.258	0.361	1.213	0.224	0.003
λ_{41}	2.399	0.677	2.791	0.818	2.796	0.520	0.002
τ_y: τ_1	19.672	0.310	20.415	0.314	20.358	0.287	0.193
τ_2	9.562	0.221	10.070	0.219	10.085	0.217	0.193
Λ_y: λ_{21}	0.876	0.214	0.885	0.219	0.848	0.141	0.001
κ: κ_1	—	—	—	—	−0.381	0.103	13.683[a]
Γ: γ_{21}	1.950	0.729	2.101	0.795	2.146	0.554	0.108
α: α_1	—	—	—	—	0.184	0.379	0.236[a]
Θ_δ: θ_{21}	0.469	0.149	0.404	0.140	0.453	0.099	0.103
ML:							
χ^2	15.655		8.756		27.542		
NFI	0.917		0.955		0.928		
NNFI	0.893		0.979		0.987		
ERLS:							
χ^2	15.148		8.572		26.858		
NFI	0.994		0.997		0.995		
NNFI	0.993		0.999		0.999		
df	7		7		24		

[a] These χ^2's are calculated by squaring the estimate/(standard error) ratio, because κ and α for the control group are set to zero.

8.A.5 and 8.A.6. A path diagram of the equal mean and covariance structure model generated by LIREL 8.12 is presented in Figure 8.3. Estimates, which are different from the previous model assuming equality of the covariance structure only, are again included in the figure.

Again, the parameter estimates from the two separate models do not appear to be significantly different from each other. The third model, which imposes an equality constraint on all the parameters listed in the table, does not give any significant LMT for the individual constraints. The overall LRT also indicates no difference between the two sets of parameters ($\chi^2 = 27.542$, df = 22). However, the level of SES (κ) is very different between the two groups, with the Head Start group having a significantly lower level than the control group. Once the SES effect is controlled for, the mean levels of cognitive ability (α) do not appear to be significantly different from each other, casting doubt on the effectiveness of the Head Start program.

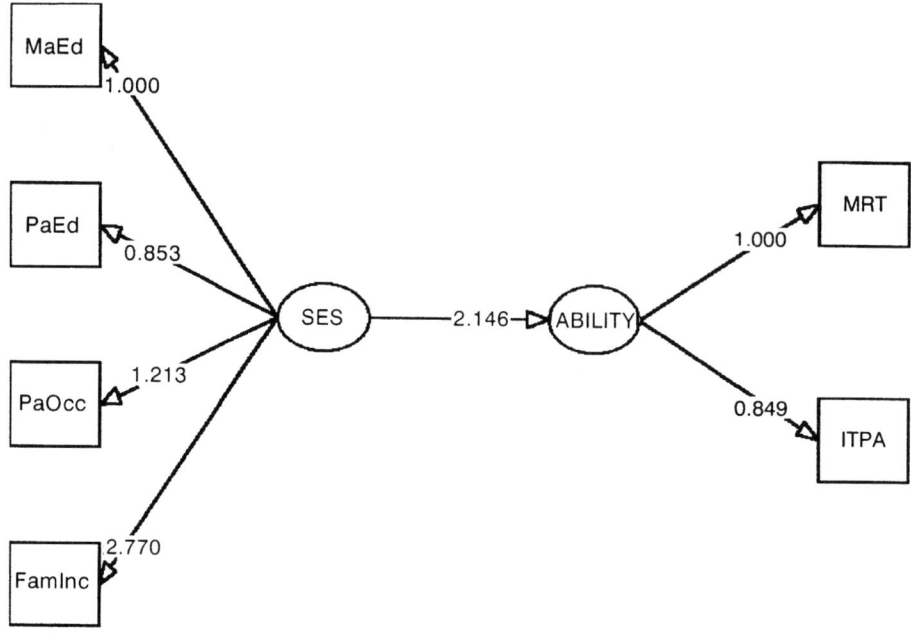

Figure 8.3 Equal mean and covariance structure for the Head Start and the control groups.

8.A EXAMPLES OF COMPUTER PROGRAM LISTINGS

To save space, the middle four lines of each correlation matrix are skipped. Refer to Tables 8.1 and 8.2 for the entries.

8.A.1 An EQS Program for Comparing Correlation Matrices

```
/title
  Head Start Two-Group Correlation Structure Model-the Head
  Start Group
/specifications
  case = 148; variables = 6; analysis = covariance; matrix =
  correlation;  method = erls; groups = 2;
/labels
  v1 = maed; v2 = paed; v3 = paocc; v4 = faminc; v5 = MRT; v6 = ITPA;
  f1 = SES; f2 = cogable;
/equations
  v1 = *f1;
  v2 = *f2;
  v3 = *f3;
  v4 = *f4;
```

```
    v5 = *f5;
    v6 = *f6;
/variances
    f1 to f6 = 1;
/covariances
    f1 to f6 = *;
/matrix
    1.000
    ...
    0.270 0.122 0.251 0.198 0.664 1.000
/standard deviations
    1.332 1.281 1.075 2.648 3.764 2.677
/end
/title
    Head Start Two-Group Correlation Structure Model-the Control
    Group
/specifications
    case = 155; variables = 6; analysis = covariance; matrix =
    correlaton; method = erls;
/labels
    v1 = maed; v2 = paed; v3 = paocc; v4 = faminc; v5 = MRT; v6 = ITPA;
    f1 = SES; f2 = cogable;
/equations
    v1 = *f1;
    v2 = *f2;
    v3 = *f3;
    v4 = *f4;
    v5 = *f5;
    v6 = *f6;
/variances
    f1 to f6 = 1;
/covariances
    f1 to f6 = *;
/matrix
    1.000
    ...
    0.265 0.297 0.234 0.162 0.635 1.000
/standard deviations
    1.360 1.195 1.193 3.239 3.900 2.719
/constraints
    (1,f1,f2) = (2,f1,f2)
    (1,f1,f3) = (2,f1,f3)
    (1,f1,f4) = (2,f1,f4)
    (1,f1,f5) = (2,f1,f5)
    (1,f1,f6) = (2,f1,f6)
    (1,f2,f3) = (2,f2,f3)
    (1,f2,f4) = (2,f2,f4)
    (1,f2,f5) = (2,f2,f5)
    (1,f2,f6) = (2,f2,f6)
```

```
  (1,f3,f4) = (2,f3,f4)
  (1,f3,f5) = (2,f3,f5)
  (1,f3,f6) = (2,f3,f6)
  (1,f4,f5) = (2,f4,f5)
  (1,f4,f6) = (2,f4,f6)
  (1,f5,f6) = (2,f5,f6)
/lmtest
/end
```

8.A.2 A LISREL Program for Comparing Correlation Matrices

```
Head Start Two-Group Correlation Structure Model-the Head
  Start Group
DA NO = 148 NI = 6 NG = 2
KM *
  1.000
  ...
  0.270 0.122 0.251 0.198 0.664 1.000
SD *
  1.332 1.281 1.075 2.648 3.764 2.677
LA *
  maed paed paocc faminc MRT ITPA
MO NX = 6 NK = 6 TD = FI
FR LX 1 1 LX 2 2 LX 3 3 LX 4 4 LX 5 5 LX 6 6
OU SE TV ND = 3

Head Start Two-Group Correlation Structure Model-the Control
  Group
DA NO = 155
KM *
  1.000
  ...
  0.265 0.297 0.234 0.162 0.635 1.000
SD *
  1.360 1.195 1.193 3.239 3.900 2.719
LA *
  maed paed paocc faminc MRT ITPA
MO NX = 6 NK = 6 TD = FI PH = IN
FR LX 1 1 LX 2 2 LX 3 3 LX 4 4 LX 5 5 LX 6 6
Path Diagram
OU SE TV ND = 3
```

8.A.3 An EQS Program for Comparing Covariance Structures

```
/title
  Head Start Two-Group Covariance Structure Model-the Head
  Start Group
/specifications
  case = 148; variables = 6; analysis = covariance; matrix =
  correlaton; method = erls; groups = 2;
```

```
/labels
  v1 = maed; v2 = paed; v3 = paocc; v4 = faminc; v5 = MRT; v6 = ITPA;
  f1 = SES; f2 = cogable;
/equations
  v1 = f1 + e1;
  v2 = *f1 + e2;
  v3 = *f1 + e3;
  v4 = *f1 + e4;
  v5 = f2 + e5;
  v6 = *f2 + e6;
  f2 = *f1 + d1;
/variances
  e1 to e6 = *;
  d1 = *;
/covariances
  e1,e2 = *;
/print effect = yes; /matrix
  1.000
  ...
  0.270 0.122 0.251 0.198 0.664 1.000
/standard deviations
  1.332 1.281 1.075 2.648 3.764 2.677
/end
/title
  Head Start Two-Group Covariance Structure Model-the Control
  Group
/specifications
  case = 155; variables = 6; analysis = covariance; matrix =
  correlaton method = erls;
/labels
  v1 = maed; v2 = paed; v3 = paocc; v4 = faminc; v5 = MRT; v6 = ITPA;
  f1 = SES; f2 = cogable;
/equations
  v1 = f1 + e1;
  v2 = *f1 + e2;
  v3 = *f1 + e3;
  v4 = *f1 + e4;
  v5 = f2 + e5;
  v6 = *f2 + e6;
  f2 = *f1 + d1;
/variances
  e1 to e6 = *;
  d1 = *;
/covariances
  e1,e2 = *;
/print
  effect = yes;
/matrix
  1.000
  ...
  0.265 0.297 0.234 0.162 0.635 1.000
```

```
/standard deviations
  1.360 1.195 1.193 3.239 3.900 2.719
/constraints
  (1,v2,f1) = (2,v2,f1)
  (1,v3,f1) = (2,v3,f1)
  (1,v4,f1) = (2,v4,f1)
  (1,v6,f2) = (2,v6,f2)
  (1,f2,f1) = (2,f2,f1)
  (1,e1,e2) = (2,e1,e2)
/lmtest
/end
```

8.A.4 A LISREL Program for Comparing Covariance Structures

```
Head Start Two-Group Covariance Structure Model-the Head Start
  Group
DA NO = 148 NI = 6 NG = 2
KM *
  1.000
  ...
  0.270 0.122 0.251 0.198 0.664 1.000
SD *
  1.332 1.281 1.075 2.648 3.764 2.677
LA *
  maed paed paocc faminc MRT ITPA
SE 5 6 1 2 3 4
MO NX = 4 NK = 1 NY = 2 NE = 1 LX = FR LY = FR TD = SY
LK SES
LE ABILITY
FI LX 1 1 LY 1 1
VA 1 LX 1 1 LY 1 1 FR TD 2 1
OU SE TV ND = 3

Head Start Two-Group Covariance Structure Model-the Control
  Group
DA NO = 155
KM *
  1.000
  ...
  0.265 0.297 0.234 0.162 0.635 1.000
SD *
  1.360 1.195 1.193 3.239 3.900 2.719
LA *
  maed paed paocc faminc MRT ITPA
SE 5 6 1 2 3 4
MO NX = 4 NK = 1 NY = 2 NE = 1 LX = IN LY = IN GA = IN TD = SP
LK SES
LE ABILITY
EQ TD(1,2,1) = TD(2,2,1)
Path Diagram
OU SE TV ND = 3
```

8.A.5 An EQS Program for Comparing Mean and Covariance Structures

```
/title
  Head Start Two-Group Mean Covariance Structure Model-the
  Head Start Group
/specifications
  case = 148; variables = 6; analysis = moment; matrix = correlaton;
  method = erls; groups = 2;
/labels
  v1 = maed; v2 = paed; v3 = paocc; v4 = faminc; v5 = MRT; v6 = ITPA;
  f1 = SES; f2 = cogable;
/equations
  v1 = 3*v999 + f1 + e1;
  v2 = 3*v999 + 1*f1 + e2;
  v3 = 2*v999 + 1*f1 + e3;
  v4 = 5*v999 + 2*f1 + e4;
  v5 = 20*v999 + f2 + e5;
  v6 = 10*v999 + 1*f2 + e6;
  f1 = *v999 + d1;
  f2 = *v999 + 2*f1 + d2;
/variances
  e1 to e6 = 1.5*;
  d1 to d2 = 0.5*;
/covariances
  e1,e2 = *;
/print effect = yes; /matrix
  1.000
  ...
  0.270 0.122 0.251 0.198 0.664 1.000
/standard deviations
  1.332 1.281 1.075 2.648 3.764 2.677
/means
  3.520 3.081 2.088 5.358 19.672 9.562
/end /title
  Head Start Two-Group Mean Covariance Structure Model-the
  Control Group
/specifications
  case = 155; variables = 6; analysis = moment; matrix = correlaton;
  method = erls;
/labels
  v1 = maed; v2 = paed; v3 = paocc; v4 = faminc; v5 = MRT; v6 = ITPA;
  f1 = SES; f2 = cogable;
/equations
  v1 = 3*v999 + f1 + e1;
  v2 = 3*v999 + 1*f1 + e2;
  v3 = 2*v999 + 1*f1 + e3;
  v4 = 5*v999 + 2*f1 + e4;
  v5 = 20*v999 + f2 + e5;
  v6 = 10*v999 + 1*f2 + e6;
```

```
  f1 = 0v999 +d1;
  f2 = 0v999 + 2*f1 + d2;
/variances
  e1 to e6 = 1.5*;
  d1 to d2 = 0.5*;
/covariances e1,e2 = *; /print effect = yes; /matrix
  1.000
  ...
  0.265 0.297 0.234 0.162 0.635 1.000
/standard deviations
  1.360 1.195 1.193 3.239 3.900 2.719
/means
  3.839 3.291 2.600 6.435 20.415 10.070
/constraints
  (1,v1,v999) = (2,v1,v999)
  (1,v2,v999) = (2,v2,v999)
  (1,v3,v999) = (2,v3,v999)
  (1,v4,v999) = (2,v4,v999)
  (1,v5,v999) = (2,v5,v999)
  (1,v6,v999) = (2,v6,v999)
  (1,v2,f1) = (2,v2,f1)
  (1,v3,f1) = (2,v3,f1)
  (1,v4,f1) = (2,v4,f1)
  (1,v6,f2) = (2,v6,f2)
  (1,f2,f1) = (2,f2,f1)
  (1,e1,e2) = (2,e1,e2)
/lmtest
/end
```

8.A.6 A LISREL Program for Comparing Mean and Covariance Structures

```
Head Start Two-Group Mean Covariance Structure Model-the Head
  Start Group
DA NO = 148 NI = 6 NG = 2
KM *
  1.000
  ...
  0.270 0.122 0.251 0.198 0.664 1.000
SD *
  1.332 1.281 1.075 2.648 3.764 2.677
ME *
  3.520 3.081 2.088 5.358 19.672 9.562
LA *
  maed paed paocc faminc MRT ITPA
SE 5 6 1 2 3 4
MO NX = 4 NK = 1 NY = 2 NE = 1 LX = FR LY = FR TX = FR TY = FR AL = FR
  KA = FR TD = SY
LK SES
```

```
LE ABILITY FI LX 1 1 LY 1 1
VA 1 LX 1 1 LY 1 1
FR TD 2 1
OU SE TV ND = 3
```

Head Start Two-Group Mean Covariance Structure Model-the
  Control Group
```
DA NO = 155
KM *
 1.000
 ...
 0.265 0.297 0.234 0.162 0.635 1.000
SD *
 1.360 1.195 1.193 3.239 3.900 2.719
ME *
 3.839 3.291 2.600 6.435 20.415 10.070
LA *
 maed paed paocc faminc MRT ITPA
SE 5 6 1 2 3 4
MO NX = 4 NK = 1 NY = 2 NE = 1 LX = IN LY = IN GA = IN TX = IN TY = IN
  AL = FI KA = FI TD = SP
LK SES
LE ABILITY
EQ TD(1,2,1) = TD(2,2,1)
Path Diagram
OU SE TV ND = 3
```

CHAPTER 9

Comparison with Categorical Latent Variables

9.1 INTRODUCTION

The last chapter dealt with statistical group comparisons in SEM. When latent variables are present in a structural equation model, they are assumed to be continuous, and are typically assumed to follow the multivariate normal distribution just as what underlies the observed variables. However, just as observed variables can be continuous and categorical, latent variables can be both as well. Here we define a variable to be *continuous* when it is measured on a metric or interval scale, and *categorical* when it is measured on an ordinal or nominal basis. Latent variables represent concepts that we can define continuously or discretely. For example, SES and cognitive ability are studied as continuous latent variables in Chapter 8. However, a latent variable representing the concept of "expert" or "heroic behavior" may be better kept as categorical.

As many concepts in the social and behavioral sciences are difficult to measure directly, indicator or manifest variables are often observed as indirect measures of the concepts that are latent. The development in SEM is just one aspect of the larger topic on latent structure models, a general term for statistical models with latent variables. Early works on the categorical variant of the latent structure models are summarized and discussed in Lazarsfeld and Henry (1968). Among the many recent treatments, Bartholomew and Knott (1999) and Vermunt (1997) gave excellent introductions not only to categorical latent structure models but also to the relation between them and models with continuous latent variables as well as other related methods such as causal loglinear models with latent variables.

Because both the latent and the observed variables can be measured as continuous or categorical, one may classify latent structure models according to the measurement level of the variables (Bartholomew and Knott 1999). A cross-classification of the measurement level of the latent and the observed variables produces a 2 × 2 table (Table 9.1).

Table 9.1 A Typology of Latent Structure Models

Observed Variable	Latent Variable	
	Categorical	Continuous
Categorical	Latent class model	$\left\{ \begin{array}{l} \text{Latent trait model} \\ \text{Factor analytic model} \end{array} \right\}$
Continuous	$\left\{ \begin{array}{l} \text{Latent class model} \\ \text{Latent profile model} \end{array} \right\}$	Factor analytic model

Thus, in factor analysis, which is a special case of SEM, one of more continuous latent variables are assumed to underlie a set of continuous indicators. In a latent trait model, which is based on item response theory, a continuous latent variable is measured by a set of categorical observed variables. The same combination is true for a measurement (or factor) model in the SEM tradition when the observed variables are categorical. In a latent profile model (and a latent class model with linear restrictions), the latent variable is categorical while the observed indicators are continuous. Finally, when both the latent and the observed variables are categorical, one obtain a typical latent class model. Note that even though we say factor analysis is a special case of SEM, recent advances in applied statistical methodology have made the boundaries less clear-cut. One may, for example, incorporate a latent class model or a Rasch model as the measurement model in a general structural equation framework to study causal effects of latent variables.

In this chapter we focus on the first column of Table 9.1, that is, latent class and related models for categorical latent variables, although we also briefly discuss latent trait models, because they are also based on categorical data and hence often discussed together with latent class models. In Section 9.2 we introduce latent class models (LCMs); in Section 9.3 we discuss latent trait models (LTMs). The latent profile model (LPM) is examined for handling continuous observed variables in Section 9.4. We examine SEM – type causal models with categorical latent variables in Section 9.5, in preparation for a discussion of comparison with categorical latent variables in Section 9.6. Finally, in Section 9.7 we illustrate comparison with latent categorical variables. Because most of the software for categorical latent structure models are fairly specialized, an appendix lists some major software, and a second appendix lists some programs for the example.

9.2 LATENT CLASS MODELS

The LCM is presented in this section. Suppose that we have data available on K observed (or manifest) variables X_1 to X_K with indices x_1 to x_K for each of N observations; each manifest variable can be dichotomous or ordinal. Suppose that the values of any manifest variable x_k are associated with a categorical latent variable Ξ with index ξ. Here we employ uppercase for a variable, and the lowercase for its index or levels. For consistency with Chapter 8, we continue the use of the notation ξ and x though the latent variable now is categorical.

The basic equation of the classical LCM is

$$\pi_{x_1 x_2 \cdots x_K} = \sum_{\xi} \pi_{\xi x_1 x_2 \cdots x_K}, \tag{9.1}$$

where

$$\pi_{\xi x_1 x_2 \cdots x_K} = \pi_\xi \pi_{x_1 x_2 \cdots x_K | \xi} = \pi_\xi \pi_{x_1 | \xi} \pi_{x_2 | \xi} \cdots \pi_{x_K | \xi}, \tag{9.2}$$

in which π_ξ denotes the probability of belonging to latent class ξ, $\pi_{x_1 x_2 \cdots x_K}$ denotes the probability of being in cell $(\xi, x_1, x_2, \ldots, x_K)$ of the joint distribution of $\Xi X_1 X_2 \cdots X_K$; and $\pi_{x_k | \xi}$ is $P(X_k = x_k | \Xi = \xi)$, the conditional probability of being in category x_k of variable X_k, given that one belongs to latent class ξ.

Let us reuse the SES example from the previous chapter to illustrate. Suppose that SES is a categorical variable with levels of lower, middle, and upper classes, and that mother's education, father's education, father's occupation, and family income are all measured in ordered categories in X_1 to X_4 (Figure 9.1). The joint probability of belonging to a particular SES class with a particular set of values in the four manifest variables is, using (9.2),

$$\pi_{x_1 x_2 x_3 x_4} = \pi_\xi \pi_{x_1 | \xi} \pi_{x_2 | \xi} \pi_{x_3 | \xi} \pi_{x_4 | \xi}.$$

There are three properties we can summarize about the LCM. First, the latent classes are assumed to be exhaustive and mutually exclusive. The

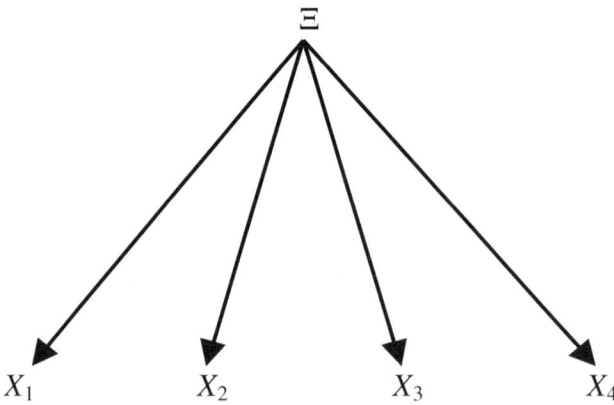

Figure 9.1 A latent class model of SES.

LATENT CLASS MODELS   145

latent SES class probabilities sum to 1. Second, the categories of each manifest variable are assumed to be exhaustive and mutually exclusive. This means that within each SES class the probabilities of the observed categories sum up to 1. Finally and most fundamentally, the LCM assumes local independence:

$$X_k | \Xi \sim \text{independent}. \tag{9.3}$$

The assumption suggests that the manifest variables are unrelated to one other once the latent variable is controlled. That is, the observed SES variables should be independent of one another within each latent SES class.

This is the classical LCM, and typically the classical parameterization as in Lazarsfeld and Henry (1968) is given in probability form as (9.2). Haberman (1979), however, demonstrated that the classical LCM of (9.2) is formally identical to the hierarchical loglinear model $\{\Xi X_1, \Xi X_2, \ldots, \Xi X_K\}$. This loglinear model can be written as

$$\ln F_{\xi x_1 x_2 \cdots x_K} = u + u_\xi^\Xi + u_{x_1}^{X_1} + u_{x_2}^{X_2} + \cdots + u_{x_K}^{X_K} + u_{\xi x_1}^{\Xi X_1} + u_{\xi x_2}^{\Xi X_2} + \cdots + u_{\xi x_K}^{\Xi X_K}, \tag{9.4}$$

where

u = grand mean effect (or intercept),
u_ξ^Ξ = first-order effect of latent variable Ξ,
$u_{x_k}^{X_k}$ = first-order effect of manifest variable X_k,
$u_{\xi x_k}^{\Xi X_k}$ = second-order or association effect of Ξ and X_k,
$F_{\xi x_1 x_2 \cdots x_K} = N \pi_{\xi x_1 x_2 \cdots x_K}$ = expected frequency of cell $(\xi, x_1, x_2, \ldots, x_K)$,

and where

$$\sum_\xi u_\xi^\Xi = \sum_{x_1} u_{x_1}^{X_1} = \cdots = \sum_{x_K} u_{x_K}^{X_K} = \sum_\xi u_{\xi x_1}^{\Xi X_1}$$

$$= \sum_{x_1} u_{\xi x_1}^{\Xi X_1} = \cdots = \sum_\xi u_{\xi x_K}^{\Xi X_K} = \sum_{x_K} u_{\xi x_1}^{\Xi X_K} = 0.$$

The usual parameter constraints are applied. The loglinear parameterization includes only the overall mean, the one-variable terms, and the two-variable interaction terms between the latent variable Ξ and the observed variables. As none of the interactions between the manifest variables are in (9.4), it can be taken as a way to specify the classical LCM with the assumption that they are conditionally independent of one another. This is the important *local independence* assumption.

The parameters in the two traditions—the loglinear and the classical—are related, of course. The conditional probabilities in (9.2), according to Haberman (1979, p. 551) and Vermunt (1997, p. 51), can be expressed in terms of the loglinear parameters from (9.4):

$$\pi_{x_k|\xi} = \frac{\exp\left(u_{x_k}^{X_k} + u_{\xi x_k}^{\Xi X_k}\right)}{\sum_{x_k} \exp\left(u_{x_k}^{X_k} + u_{\xi x_k}^{\Xi X_k}\right)}. \tag{9.5}$$

Because of the local independence assumption, the estimated relationships between the latent and the manifest variables from a separate logit model for each observed variable should be identical to those same relationships from a loglinear model for the contingency table with the hierarchical marginals $\{\Xi X_1, \Xi X_2, \ldots, \Xi X_K\}$.

So far we have only discussed the unrestricted LCM. Restrictions, as in other statistical models, can be applied. For example, in the classical parameterization, it is typical to apply fixed-value and equality restrictions to the latent and conditional probabilities. In the loglinear parameterization, on the other hand, it is common to apply to the loglinear parameters certain restrictions such as equal effects of the manifest variables, linear-by-linear associations, row effects, column effects, and row and column effects. Less common are nonlinear or inequality restrictions. By applying particular kinds of inequality restrictions on the conditional response probabilities, one may obtain an ordinal latent class model (Croon 1990; Vermunt 1997). This may prove to be particularly useful when all the manifest variables are also ordinal.

In spite of the importance of the local independence assumption, the LCM may benefit from relaxing it. Given the latent variable, certain indicators may still relate to each other. Using the SES example, mother's and father's education can be related even though the latent categorical SES is controlled. This calls for the local dependence model. Its application using the SES example can be illustrated by Figure 9.2. The loglinear LCM can be easily used to specify such models. Hagenaars (1988) demonstrated the specification of the local dependence model. For the SES example, this is a loglinear model of $\{\Xi X_1, \Xi X_2, \Xi X_3, \Xi X_4, X_1 X_2\}$. That is, there exists a direct association between X_1 and X_2, or mother's and father's education. In the classical parameterization, the model can be written as

$$\pi_{\xi x_1 x_2 x_3 x_4} = \pi_\xi \pi_{x_1 x_2|\xi} \pi_{x_3|\xi} \pi_{x_4|\xi}, \tag{9.6}$$

where $\pi_{x_1 x_2|\xi}$ is restricted by way of a no-three-variable-interaction loglinear model. The model resembles the measurement model with correlated error terms in Chapter 8.

LATENT TRAIT MODELS

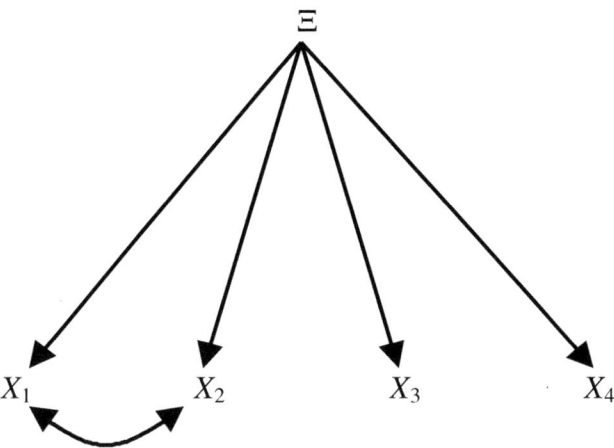

Figure 9.2 A local dependence model of SES.

9.3 LATENT TRAIT MODELS

In LTMs the latent variable is called a latent trait, and the relation between it and the manifest response is analyzed by modeling the probability that an observation with latent score ξ will respond in category x_k of item X_k. Thus, the focus is on how the conditional probability $\pi_{x_k|\xi}$ is modeled. In item response theory, the latent trait is regarded as a continuous variable.

When both observations and items are located on the same unidimensional continuum, the probability $\pi_{x_k|\xi}$ increases monotonically with the difference $\xi - x_k$. The fashion in which the item response function is modeled using certain cumulative distribution functions gives rise to two variants of the LTM. One is called Gaussian or *normal ogive* (ogive referring to the characteristic S shape of the item response function):

$$\pi_{x_k|\xi} = \Phi(\xi - x_k), \tag{9.7}$$

where Φ is the cumulative standard normal distribution function. Such a model derives from the assumption of normally distributed measurement error. The Gaussian version of the LTM can be estimated using SEM software that allows categorical variables. The other variant consists of the logistic-ogive and Rasch models, which model the conditional probability by

$$\pi_{x_k|\xi} = \frac{\exp(\xi - x_k)}{1 + \exp(\xi - x_k)}. \tag{9.8}$$

It appears that (9.8) shares striking resemblance to (9.5) for the LCM. The importance difference between the two lies in the continuity of the latent dimension of (9.8) and the discrete nature of the LCM. This distinction determines whether the function in question follows a cumulative logistic distribution. The majority of the literature on item response theory has focused on this type of LTM.

The specifications above define a latent variable model with a continuous trait. Restrictions can be imposed to require treating the indicators as ordered variables and linearizing the interaction between the indicators and the latent variable. This entails having a set of fixed category values of the manifest variables and fixed *node-points* of the latent variable, according to a scaling parameter a_k. The LTM with such restrictions is known as the discrete LTM.

When the latent trait is discretized, only a certain node points on the latent continuum play a part in the formalization of the LTM. Bock (1972) generalized the LTM to

$$\pi_{x_k|\xi} = \frac{\exp(a_k \xi + c_k)}{\sum_m \exp(a_k \xi + c_k)}, \qquad (9.9)$$

where $c_k = -a_k \times x_k$ and the summation is over the categories of item x_k. It is assumed that, for level m of M response categories in X_k, the following inequality assumption holds:

$$a_{k1} < a_{k2} < \cdots < a_{k\,M-1} < a_{kM}. \qquad (9.10)$$

A model like this can be estimated with several of the software packages in Section 9.A. However, in order for a latent trait specification to be sensible, it is best to have a large number of items. Latent trait models have recently been extended to a class called generalized latent trait models, which can analyze manifest variables with different distributions in the exponential family (Moustaki and Knott 2000).

9.4 LATENT VARIABLE MODELS FOR CONTINUOUS INDICATORS

The LCM deals with binary or polytomous data by specifying a suitable form for the conditional distribution of x_k; the binomial and multinomial distributions are the natural choices. When the observed variables are continuous, we are concerned with a suitable distributional form for the joint distribution (Bartholomew and Knott 1999):

$$f(x_1, x_2, \ldots, x_K) = \sum_\xi \xi \prod_{k=1}^{K} h_k(x_k|\xi), \qquad (9.11)$$

where $h_k(x_k|\xi)$ is the conditional distribution of x_k given class ξ. Models following this specification have been termed latent profile models.

Oftentimes we have no information on the form of $h_k(x_k|\xi)$. When that happens, an inspection of the marginal distribution

$$f(x_k) = \sum_\xi \xi h_k(x_k|\xi)$$

should help. For example, a skewed distribution would give evidence against a normal mixture of any number of components, whereas a clear bimodal distribution would indicate a two-class normal mixture. To estimate a latent profile model like this, Lazarsfeld and Henry (1968) proposed to use the method of moments; later treatments relied on the EM method.

Just as with applying linear restrictions for modeling a latent trait, loglinear models can prove to be the most flexible for modeling the relationship between a categorical latent variable and some metric indicators as well. As Heinen (1993) pointed out, though not widely applied, loglinear models with row or column association can handle the situation. Specifically, consider all the marginals formed by the categorical latent variable and any of the metric manifest variables as a series of two-way classifications. Depending on whether the metric observed variable is the row or the column, a column- or row-effect association model with linear restrictions applied on the other dimension containing the observed variable is the obvious choice. The model is restricted in that for each level ξ of Ξ the total number of interaction parameters is reduced from the number of categories in X_k minus one to only the single parameter $u_\xi^{\Xi X_k}(x_k - \bar{x}_k)$, with the x_k's being the metric values of X_k. Therefore, the two-way interaction terms in (9.4) are replaced by

$$u_{\xi x_k}^{\Xi X_k} = u_\xi^{\Xi X_k}(x_k - \bar{x}_k) \tag{9.12}$$

where $k = 1, 2, \ldots, K$. This shows the flexibility of the loglinear formulation, which can be made to generate a LCM, or a LTM, and a LPM by simply applying restrictions on the loglinear parameters.

9.5 CAUSAL MODELS WITH CATEGORICAL LATENT VARIABLES

In this section we consider two extensions to the latent variable models discussed so far. First, we allow for more than one latent variable in a latent structure model. Then we extend it to causal structural models along the lines of Goodman's (1973) modified path models for loglinear analysis. Furthermore, Clogg (1981) used LCMs with external variables to specify a multiple-indicator multiple-cause (MIMIC) model for categorical data. Hagenaars (1993) gave an accessible introduction to loglinear models with latent variables.

Analogously to a factor analytic model with more than one factor, LCMs can include more than one latent variable (Haberman 1979). Such models

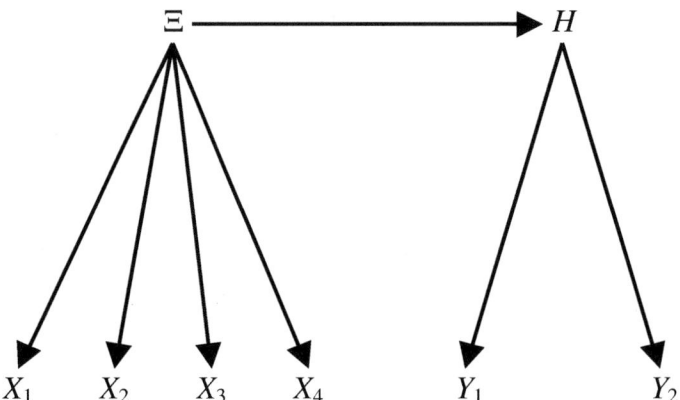

Figure 9.3 A casual model with latent variables for SES and cognitive ability.

can be specified either by imposing equality restrictions on the conditional probabilities or by formulating a loglinear model.

Let us continue with the SES example. By using categorized versions of the variables, we may revisit the Head Start example from Chapter 8. With X_1 to X_4 indicating the four SES variables and Y_1 and Y_2 representing MRT and ITPA, we use H to denote the latent variable of cognitive ability with its levels referred to by η. See Figure 9.3 for a graphic representation of the model. A categorical version of the model tested in SEM can be formed, and this results in the loglinear latent class model $\{\Xi H, \Xi X_1, \Xi X_2, \Xi X_3, \Xi X_4, HY_1, HY_2\}$, which can be expressed as

$$\ln F_{\xi \eta x_1 x_2 x_3 x_4 y_1 y_2} = u + u_\xi^\Xi + u_\eta^H + u_{x_1}^{X_1} + u_{x_2}^{X_2} + u_{x_3}^{X_3} + u_{x_4}^{X_4} + u_{y_1}^{Y_1} + u_{y_1}^{Y_1}$$
$$+ u_{\xi\eta}^{\Xi H} + u_{\xi x_1}^{\Xi X_1} + u_{\xi x_2}^{\Xi X_2} + u_{\xi x_3}^{\Xi X_3} + u_{\xi x_4}^{\Xi X_4} + u_{\eta y_1}^{HY_1} + u_{\eta y_2}^{HY_2},$$
(9.13)

where again no higher than second-order terms are included. Additional restrictions can be placed on the loglinear parameters to deal with continuous indicators, for example. In a model with three or more latent variables, a model with no third-order or higher interaction can be specified.

The same model can be expressed in conditional probabilities

$$\pi_{\xi x_1 x_2 x_3 x_4 \eta y_1 y_2} = \pi_\xi \pi_{x_1|\xi} \pi_{x_2|\xi} \pi_{x_3|\xi} \pi_{x_4|\xi} \pi_{\eta|\xi} \pi_{y_1|\eta} \pi_{y_2|\eta}, \quad (9.14)$$

where a new type of conditional probability $\pi_{\eta|\xi}$ that involves latent variables only is included. The two model specifications are equivalent.

One may quickly discover that, unlike in SEM, interaction effects of two latent variables on a third can be estimated with ease, specified either in classical probability forms or as loglinear parameters. For example, we may have two exogenous latent variables Ξ_1 and Ξ_2, and one endogenous latent variable H, each of which is indicated by two manifest variables. The part of the loglinear model involving the latent variables is $\{\Xi_1\Xi_2, \Xi_1H, \Xi_2H, \Xi_1\Xi_2H\}$. The subtables (or marginal tables) that involves the manifest variables are omitted in the presentation for brevity. Equivalently, the conditional probabilities include an additional term $\pi_{\eta|\xi_1\xi_2}$. The computer programs PANMARK, LCAT, LEM, and Latent GOLD can all estimate causal models with categorical latent variables such as this, and LAT and NEWTON can also estimate modified path model, though with some extra work.

9.6 COMPARISON WITH CATEGORICAL LATENT VARIABLES

Analogously to comparison in terms of the structural model and the measurement model in SEM, one may be interested in a comparison, between multiple groups or populations of the latent distributions, of the relationships between the latent variables and the manifest variables, or both. Clogg and Goodman (1984, 1985) made an important contribution by presenting a simultaneous latent structure model for multiple populations by using the classic parameterization of the LCM. Their approach involves including in the model a group variable, which may affect the latent distribution and the conditional probabilities. The approach is presented later in the section.

It can be demonstrated that the loglinear model is a member of the GLM. Assuming a Poisson distribution, the relationship between the Poisson and the exponential family was shown in Chapter 6. As a member of the GLM, the model is amenable to the standard method for comparing multiple groups discussed there. What follows is a comparison scheme specifically organized for models with categorical latent variables. Although typically the loglinear model assumes a Poisson distribution, it can also use the multinomial or the product-multinomial as the basis of distribution. These three sampling models are used for the contingency table:

Poisson. We consider individual Poisson processes observed over a fixed period of time, one for each cell of the table, with no prior knowledge of N, the total number of observations.

Multinomial. We obtain a fixed N and cross-classify each observation according to its values for the variables in the contingency table.

Product-Multinomial. For each category of the row (or column, layer, etc.) variable, we take a multinomial sample of size $n_{i+\ ..}$ ($n_{+j+\ ..}$, etc.) and cross-classify each member of the sample according to its value for the other variables in the table.

This raises the issue of whether the sampling models underlying the multiple groups are identical. Because the sampling assumptions lead to the same expected frequencies and the fitting of a loglinear model, the comparison of sampling models across groups is not really necessary. It is, however, necessary to compare the actual sampled frequencies in the table across groups, as they have different patterns.

9.6.1 Comparing Sampling Distributions

Almost invariably models with categorical latent variables are based on contingency tables. Loglinear models are sensitive to the frequency distributions in the table. Different groups have quite different frequencies for the same cells. The convention is that the cells in a contingency table preferably contain at least one observation. Problems may occur when one group has numerous sampling zeros in a number of cells while the other groups have a sizable number of observations in the same cells. It may be useful, therefore, to compare the patterns of the sampling distributions, especially in terms of sampling zeros, before moving on to comparing model probabilities or parameters.

The null hypothesis is to assume the groups have the same pattern of sampling zeros:

$$H_0: \quad F_g^{(0)} = F_h^{(0)},$$

where the superscript (0) denotes a pattern of sampling zeros for the expected frequencies of group g under a certain model. To test such a hypothesis, we may treat the sampling zeros as if they were structural zeros, and analyze the data under a null model (such as a grand-mean-only model) for each group by contrasting the zeros cells (and the nonzero cells) between the groups. Likelihood ratio statistics can then be compared between pairs of groups to conduct a LRT, though tests can be performed more easily on a single loglinear parameter that represents the zero patterns. A significant test may suggest a different zero pattern.

9.6.2 Comparing Types and Patterns of Association between Variables

In loglinear analysis, the association in a two-way table can be modeled in a number of ways, such as second-order interaction, linear-by-linear association, row-effect association, column-effect association, and log-multiplicative association. Similarly, the association existing in the subtables formed by a latent variable on one hand and a manifest variable on the other or by two latent variables may be different across groups. For such scenarios, we have the null hypothesis of

$$H_0: \quad A_{g \equiv X} = A_{h \equiv X} \ \wedge \ A_{gHY} = A_{hHY} \ \wedge \ A_{g \equiv H} = A_{h \equiv H},$$

where $A_{g \equiv X}$ stands for the type of association assumed between Ξ and X for group g, for example. The comparison can be performed by making different types of association assumption for different groups and testing with the LRT.

Sometimes the association among the manifest variables of one group has a different pattern from that of another group, so that the association may give rise to a different number of latent classes in different groups. The null hypothesis then becomes

$$H_0: \dim(\Xi_g) = \dim(\Xi_h),$$

where $\dim(\Xi_g)$ denotes the dimension of the latent variable Ξ for group g. This hypothesis can be tested alone or in conjunction with the hypothesis of the type of association between variables. Practically, it is a useful hypothesis to test.

9.6.3 Comparing Conditional Structure and Response Probabilities

Groups that do not differ in their basic data patterns such as sampling zeros may differ in latent structures. However, groups that do differ in sampling zeros do not necessarily differ much in latent structures. Patterns of variable association account for most of the differences. As in the comparison of model parameters in SEM, the data analyst may be interested in the model conditional probabilities in the form of either a manifest variable conditional on a latent variable, a latent variable conditional on another latent variable, or a latent variable directly on an observed external variable as in the case of the MIMIC model:

$$H_{\pi_{x|\xi}, \pi_{y|\eta}, \pi_{\eta|\xi}, \pi_{\eta|x}}:$$

$$\pi_{g,x|\xi} = \pi_{h,x|\xi} \wedge \pi_{g,y|\eta} = \pi_{h,y|\eta} \wedge \pi_{g,\eta|\xi} = \pi_{h,\eta|\xi} \wedge \pi_{g,\eta|x} = \pi_{h,\eta|x}$$

where the subscript g is separated from the conditional probability by a comma to avoid confusion with joint conditional probability. Here testing a null hypothesis of $H_{\pi_{x|\xi}, \pi_{y|\eta}}$ is equivalent to testing equality in the measurement model, and a test of a null hypothesis of $H_{\pi_{\eta|\xi}, \pi_{\eta|x}}$ is equivalent to testing equality in the structural model, if we may use SEM terminology. The LRT applies readily to such tests. It is important to note that a null hypothesis like the above can be phrased in terms of loglinear parameters, even though the expression is more concise in probability form. Depending on the software, the researcher may have to express the equality test in terms of probability or loglinear parameters, but not both.

9.6.4 Comparing Latent Distributions and Conditional Probabilities

Somewhat analogous to comparing structured latent means is comparing distributions across latent classes. It is likely that a researcher will be interested in comparing the compositions of the latent classes instead of the conditional probabilities. For example, is there a difference between the probability of a child from the Head Start group belonging to the lower SES class and the probability of a child from the control group belonging to it? Is there a difference in the probability of membership in the higher cognitive ability class between the same two children? Under such research circumstances, one should test the null hypothesis

$$H_{\pi_\xi, \pi_\eta}: \quad \pi_{g,\xi} = \pi_{h,\xi} \wedge \pi_{g,\eta} = \pi_{h,\eta}.$$

It is possible that another researcher may be interested in both the latent distributions and the conditional probabilities. The null hypothesis then becomes

$$H_{\pi_\xi, \pi_\eta, \pi_{x|\xi}, \pi_{y|\eta}, \pi_{\eta|\xi}, \pi_{\eta|x}}:$$

$$\pi_{g,\xi} = \pi_{h,\xi} \wedge \pi_{g,\eta} = \pi_{h,\eta} \wedge \pi_{g,x|\xi} = \pi_{h,x|\xi} \wedge \pi_{g,y|\eta} = \pi_{h,y|\eta}$$

$$\wedge \quad \pi_{g,\eta|\xi} = \pi_{h,\eta|\xi} \wedge \pi_{g,\eta|x} = \pi_{h,\eta|x}.$$

For example, a LCM for the SES example of two groups testing equality both in latent distributions and in conditional probabilities can be expressed as

$$\pi_{g\xi x_1 x_2 x_3 x_4} = \pi_g \pi_{\xi|g} \pi_{x_1|g\xi} \pi_{x_2|g\xi} \pi_{x_3|g\xi} \pi_{x_4|g\xi}, \tag{9.15}$$

where the subscript g is included in all probabilities. This model can be cast in loglinear terminology as $\{G \Xi X_1, G \Xi X_2, G \Xi X_3, G \Xi X_4\}$, where G represents the grouping variable. Naturally, as with testing equalities in SES, any reasonable combinations of the latent and the conditional probabilities can be formed as a hypothesis for a particular research task at hand. As demonstrated in the example section, the LRT can again be applied readily.

9.7 EXAMPLES

9.7.1 Comparison in Latent Class Analysis

We begin with a LCM example without any exogenous variables. The data on Catholics in America in Table 9.2 are from Liao (1989b, Table 6.1), and were taken from five years of General Social Surveys in the 1980s.

We consider a two-class LCM with the three manifest variables Attend, Pray, and Reliten. Relig16 is used as the grouping variable. These indicators can be more succinctly expressed as X_1, X_2, X_3, and G (to avoid clutter in notation we use a separate variable for grouping), and the model presented in Figure 9.4, in which the latent variable of Ξ designates the two-class Catholic membership. Prior to fitting a LCM, let us first consider the

EXAMPLES

Table 9.2 Religious Data on White Catholic Women Aged 15–54

Relig16	Reliten	Pray	Attend	Frequency
1	1	1	1	12
1	1	1	2	6
1	1	1	3	2
1	1	2	1	3
1	1	2	2	6
1	1	2	3	3
1	1	3	1	2
1	1	3	2	2
1	1	3	3	2
1	2	1	2	0
1	2	1	3	0
1	2	2	1	3
1	2	2	2	2
1	2	2	3	9
1	2	3	1	1
1	2	3	2	0
1	2	3	3	7
2	1	1	1	105
2	1	1	2	68
2	1	1	3	21
2	1	2	1	42
2	1	2	2	59
2	1	2	3	34
2	1	3	1	13
2	1	3	2	12
2	1	3	3	6
2	2	1	1	5
2	2	1	2	15
2	2	1	3	42
2	2	2	1	5
2	2	2	2	19
2	2	2	3	104
2	2	3	1	3
2	2	3	2	11
2	2	3	3	51

Relig16 is the religion in which the respondent was raised: 1 = in another religion, 2 = as a Catholic. Reliten is strength of affiliation: 1 = not strong, 2 = strong. Pray is frequency of praying: 1 = less than once a day, 2 = once a day, 3 = several times a day. Attend is frequency of attending religious services: 1 = once a year or less, 2 = more than once a year but less than every week, 3 = every week or more often.

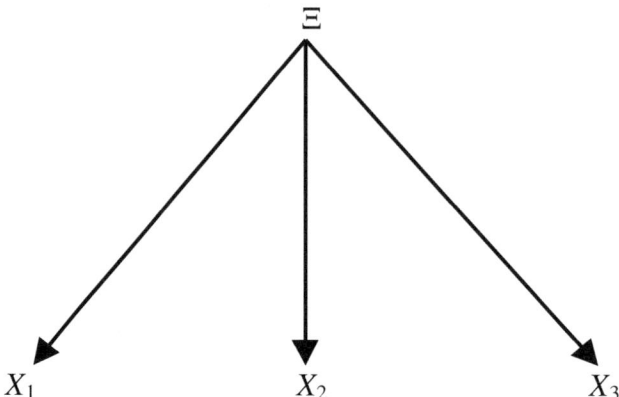

Figure 9.4 A two-class latent variable model of catholic affiliation for two groups.

sampling zero distributions, since there are four zero cells for the subtable of those raised in another religion. For this purpose we fit a loglinear model contrasting the zero cells of the Relig16 = 1 group versus the Relig16 = 2 group. We test the hypothesis

$$H_0: \quad F_1^{(0)} = F_2^{(0)}$$

that the two religious background groups have the same patterns of zeros. This can be accomplished by a design matrix with effect coding in LEM, and the program is given in Section 9.B.1.

We are not interested in the overall model fitting, which has an L^2 of 1,065 with df = 34, but in the contrast of the zeros in the two groups, represented by a loglinear parameter. The estimate is -0.239 with a standard error of 0.081, giving a Wald χ^2 of 8.670, indicating a significant difference between the two groups in terms of the zero cell patterns. We need not abandon further analysis at this point. However, we must bear the information in mind when considering more elaborate comparisons if significant differences are found.

Next, we consider various assumptions about association between the latent variable Ξ and the manifest variables X_{gk}. Typically, the two variables are treated as categorical and related by a second-order interaction $\{\Xi X_{gk}\}$ in a LCM. Let us use this model as a baseline for comparing models within the two groups. The LEM program is given in Section 9.B.2. The program is fitted to the data of both groups separately. We further consider a linear-by-linear association and a column (Ξ) effect RC-II model of the $\{\Xi X_1, \Xi X_2\}$ marginal tables, since X_1 and X_2 have three ordinal categories each. The LEM programs are given in Section 9.B.3 and 9.B.4, respectively. The model fitting information is reported in Table 9.3.

EXAMPLES

Table 9.3 Comparing Types of Association Assumptions

Association	$-2\mathcal{L}$	χ^2	L^2	Cressie–Reed	df
Group 1:					
Interaction	295.747	6.405	7.496	6.504	6
Linear × linear	299.119	11.297	10.869	10.434	8
Column RC-II	299.119	11.302	10.869	10.440	8
Group 2:					
Interaction	3,085.393	13.887	14.294	13.998	6
Linear × linear	3,090.108	18.860	19.009	18.919	8
Column RC-II	3,090.108	18.960	19.009	18.919	8

Taking the difference in χ^2 between the LCM with second-order interaction and the LCM with linear-by-linear (or column-effect RC-II) association for the marginal tables in question within each group, it is apparent that the LCM with interactions is slightly preferred for group 1, while the linear-by-linear or the column-effect RC-II association is slightly preferred for group 2. Neither of the differences, however, is significant at the 0.05 level, though both are significant at the 0.10 level. The distinction between the linear-by-linear and the column-effect RC-II association is negligible. Thus, we continue using the usual LCM with second-order interactions for the marginal tables for comparing the probabilities.

Finally, let us consider comparing the two groups in terms of classical latent class probability and conditional probabilities. We consider four latent class models. In the first, all the probabilities, conditional or not, are treated as equal between the groups:

$$\pi_{1,\xi} = \pi_{2,\xi} \wedge \pi_{1,x_k|\xi} = \pi_{2,x_k|\xi}.$$

This model assumes complete homogeneity. The negation of the model is the completely heterogeneous model:

$$\pi_{1,\xi} \neq \pi_{2,\xi} \wedge \pi_{1,x_k|\xi} \neq \pi_{2,x_k|\xi}$$

In between the two extremes, we may use the partial homogeneity assumption:

$$\pi_{1,\xi} \neq \pi_{2,\xi} \wedge \pi_{1,x_k|\xi} = \pi_{2,x_k|\xi}$$

and the partial heterogeneity assumption:

$$\pi_{1,\xi} \neq \pi_{2,\xi} \wedge \pi_{1,x_k|\xi} \neq \pi_{2,x_k|\xi} \wedge \pi_{1,\xi x_k} = \pi_{2,\xi x_k}$$

The LEM program listings are presented in Sections 9.B.5 through 9.B.8. We summarize the model fitting results in Table 9.4.

Table 9.4 Comparing Latent Class and Conditional Probabilities

Model	$-2\mathcal{L}$	χ^2	L^2	Cressie–Reed	df
Complete homogeneity	3,791.947	25.487	27.652	27.652	23
Complete heterogeneity	3,786.085	20.292	21.790	20.503	12
Partial homogeneity	3,791.871	25.516	27.576	25.614	22
Partial heterogeneity	3,788.073	21.943	23.778	22.124	17

Comparing the model fitting indices across models in the table, it is clear that neither the completely nor the partially heterogeneous model gains much by using more parameters for accommodating the second group than do the homogeneous models. The two homogeneous models are almost indistinguishable, though the Cressie–Reed statistic slightly favors the partially homogeneous model. We present probability estimates from the partially homogeneous model in Table 9.5.

Therefore, while the conditional response probabilities are held constant between the two religious background groups, the latent distributions are allowed to differ between them. However, as the table shows, the difference is not much greater than 2%; hence the closeness between the partially homogeneous and the completely homogeneous models.

9.7.2 Comparison in a Path Model with Categorical Latent Variables

For our second example, we reanalyze the data from Hagenaars (1993, Table 5.1). In his original model 4 of Table 5.2, there are two endogenous latent variables: H_1 (System Involvement) indexed by η_1, and H_2 (Protest Tolerance) indexed by η_2. The latent variable H_1 has three indicators, A (System Responsiveness), B (Ideological Level), and E (Conventional Participation), and is hypothesized to have a causal effect on the second latent variable H_2, which has two indicators C (Repression Potential) and D (Protest Approval). The latent variables are hypothesized to have two classes each. There are two external variables, T (Education) and G (Age), which are hypothesized to have a causal effect on both latent variables. A third external variable, S (Sex), is used as a grouping variable in the current example. Thus, Hagenaars's model 4 is modified for our purpose of testing equality between groups of latent distributions and model parameters; a graphical representation is given in Figure 9.5. The frequency data are presented in Section 9.B.9 of Appendix 9.B.

Our first hypothesis is one of equality between all the measurement parameters relating two categorical latent variables to their categorical indicators:

$$\pi_{1,y|\eta} = \pi_{2,y|\eta}$$

for both η's and their respective indicators (y's) for the two sex groups. Boldface symbols represent vectors of latent and observed variables, respectively.

EXAMPLES

Table 9.5 Estimates from the Partially Homogeneous Model

Probability	Estimate	S.e.
π_g		
1	0.089	0.011
2	0.911	0.011
$\pi_{\xi\mid g}$		
1|1	0.546	0.075
2|1	0.454	0.075
1|2	0.525	0.036
2|2	0.475	0.036
$\pi_{x_1\mid\xi}$		
1|1	0.510	0.035
2|1	0.419	0.029
3|1	0.071	0.033
1|2	0.030	0.017
2|2	0.160	0.027
3|2	0.810	0.035
$\pi_{x_2\mid\xi}$		
1|1	0.592	0.030
2|1	0.327	0.028
3|1	0.081	0.016
1|2	0.214	0.028
2|2	0.532	0.030
3|2	0.254	0.027
$\pi_{x_3\mid\xi}$		
1|1	0.965	0.026
2|1	0.035	0.026
1|2	0.171	0.044
2|2	0.829	0.044

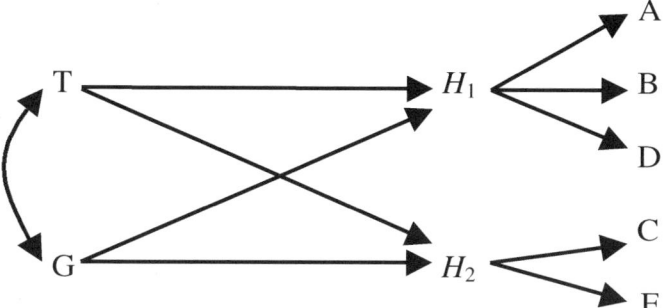

Figure 9.5 A latent path model of system involvement and protest tolerance by sex.

Next, let us test the equality constraints for the structural parameters only, that is, in SEM terminology, whether the β's and γ's differ across groups. This can be represented by the hypothesis

$$\pi_{1,\eta_2|\eta_1} = \pi_{2,\eta_2|\eta_1} \wedge \pi_{1,\eta_1|tg} = \pi_{2,\eta_1|tg} \wedge \pi_{1,\eta_2|tg} = \pi_{2,\eta_2|tg},$$

where T and G are the two external variables indexed by t and g, which does not represent groups, to keep the original notation of the example.

Finally, one may be interested in testing to see if all model parameters and latent distributions can be subjected to the equality constraints

$$\pi_{1,y|\boldsymbol{\eta}} = \pi_{2,y|\boldsymbol{\eta}} \wedge \pi_{1,\eta_2|\eta_1} = \pi_{2,\eta_2|\eta_1}$$
$$\wedge \pi_{1,\eta_1|tg} = \pi_{2,\eta_1|tg} \wedge \pi_{1,\eta_2|tg} = \pi_{2,\eta_2|tg} \wedge \pi_{1,\boldsymbol{\eta}} = \pi_{2,\boldsymbol{\eta}},$$

where again boldface $\boldsymbol{\eta}$ represents η_1 and η_2.

To test the three hypotheses, we use the LRT by comparing $-2\mathscr{L}$'s or other χ^2 type statistics between models. For this example let us use the model with all heterogeneous model parameters as the baseline. In addition to this model, we need three other models to accomplish the comparisons— a model with homogeneous structural parameters (or probabilities), a model with homogeneous measurement parameters (or probabilities), and a model with homogeneous model (both measurement and structural) parameters (or probabilities). The LEM program listings for the four models are given in Sections 9.B.10 through 9.B.13.

Among the four models, only the completely heterogeneous model fits the data moderately well, followed by the model assuming homogeneous structural parameters but heterogeneous measurement parameters. But let us focus on the comparison of the models, whose goodness-of-fit statistics are reported in Table 9.6.

With a likelihood ratio statistic of 82.266 on 24 degrees of freedom, the model with both structural and measurement parameters identical between the sex groups is significantly worse than the completely heterogeneous model, thereby rejecting the hypothesis of joint structural and measurement homogeneity. The hypothesis of homogeneous measurement parameters is also rejected at the 0.01 level, (LRT: 28.032, df = 10). However, the hypothesis of homogeneous structural parameters is rejected at the 0.05 level but retained at the 0.01 level (LRT: 20.463, df = 9). In sum, all model parameters (probabilities) are significantly different between the two sexes.

Table 9.6 Comparing Four Path Models with Categorical Latent Variables

Model	$-2\mathscr{L}$	χ^2	L^2	Cressie–Reed	df
Complete heterogeneity	12,431.527	384.578	388.687	366.620	334
Homogeneous measurement	12,459.559	411.510	416.719	393.540	344
Homogeneous structure	12,451.990	398.294	409.150	382.370	343
Complete homogeneity	12,513.793	465.904	470.953	446.882	358

In Hagenaars (1993), sex is used as an external variable. Because we have found significant differences in the model parameters between the two sexes, a stratified analysis is preferred. Although the loglinear approach is flexible enough to accommodate interaction terms, it is much easier and conceptually clearer to use a stratified model, because not only the structural parameters but also the measurement parameters are significantly different between the groups.

9.A SOFTWARE FOR CATEGORICAL LATENT VARIABLES

There are many specialized software categorical latent variable models. This is not a complete listing, and the inclusion of a program is simply based on awareness and at least some familiarity with the program. For general-purpose software, we also include an S-plus function and a SAS macro for LCMs on the list.

9.A.1 MLLSA

MLLSA (Maximum Likelihood Latent Structure Analysis) was original written by the late Clifford C. Clogg for the IBM mainframe computer. An updated version (MLLSA 4.0) for PCs (DOS) is included by Scott Eliason of the University of Minnesota as one of several programs for categorical data analysis such as hierarchical and nonhierarchical loglinear models, association models, LCM, and the purging method. The package is known as the Categorical Data Analysis System (CDAS). Either CDAS or MLLSA can be downloaded from Scott Eliason's homepage: http://www.soc.umn.edu/~eliason/CDAS.htm.

MLLSA once was the major software in use, but now there are many competitors with more flexibility and capabilities. It is easy to use, and the EM algorithm used ensures fast convergence.

9.A.2 LAT

Developed about the same time as MLLSA by Shelby Haberman of Northwestern University, LAT reparameterizes the LCM and employs a Newton–Raphson algorithm for estimation. Like its successors NEWTON and DNEWTON, LAT requires setting up a problem in terms its design matrix. This process can be painstaking for some, but can mean extra flexibility for others. For further information, contact Shelby Haberman (shelby@gibbs.stats.nwu.edu).

9.A.3 PANMARK

Written by Frank van de Pol, Rolf Langeheine, and W. de Jong, this is a flexible PC DOS program based on the EM algorithm for categorical latent structure models as well as Markov chain models. The program is marketed by ProGAMMA, accessible at http://members.nbci.com/jsuebersax/soft.htm. This user-friendly program may make users estimating larger models happier because of its feature of using multiple sets of starting values to avoid local maxima.

9.A.4 LCAG

Written by Jacques Hagenaars of Tilburg University, LCAG is a flexible program that permits local dependence models. One major advantage of the program is that the user can specify loglinear models among the latent variables. The program can be obtained from its author (jacques.a.hagenaars@kub.nl).

9.A.5 LEM

LEM is an excellent PC program written by Jeroen Vermunt of Tilburg University. It has the great flexibility for estimating a good variety of models, including latent class, latent trait, local dependence, event history, association, and causal path models with the loglinear formulation. It also can estimate models with different missing mechanisms and test parameter constraints. Its output includes both loglinear estimates and classic latent class probability estimates. The program has a DOS and a Windows 95 version, and is very user friendly, with good examples provided. The program can be downloaded from the location http://www.kub.nl/faculteiten/fsw/organisatie/departementen/mto/software2.html.

9.A.6 Latent GOLD

In September 2000 Statistical Innovations released Latent GOLD, a program resulting from two years' collaboration of Jay Magidson and Jeroen Vermunt. The software, developed out of the successful LEM, includes many of LEM flexible features. A state-of-the-art feature allows the user to apply Bayes constants in estimation. These constants define a prior distribution for probability parameter estimates so that values closer to the boundaries 0 and 1 are considered less likely, thereby providing a new solution for the problem of local maxima. A demo version of the software can be downloaded from http://latentclass.com/.

Another important difference between Latent GOLD and most other software is its fully developed Windows capabilities, which resembles those of SPSS. In fact, it can directly read SPSS data as well as ASCII data.

9.A.7 Mplus

Bengt Muthén's LISCOMP was the only software that could handle categorical data in the late 1980s. In recent years he and Linda Muthén extended LISCOMP to a more general program to handle mixtures of latent distributions, permitting the inclusion of both categorical and continuous variables. The program can also handle multilevel analyses. For more information, visit http://www.statmodel.com/.

9.A.8 LATCLASS, TWOMISS, and POLYMISS

These programs form the supporting software for Bartholomew and Knott's (1999) book on latent variables, and can be downloaded for free from the site http://www.arnoldpublishers.com/support/lvmfa2.htm.

LATCLASS fits a standard latent class model (with up to 15 classes and 30 items) using an EM algorithm. TWOMISS fits a one- or two-factor latent trait model with a logit–probit response function to binary data when observations may be missing, and POLYMISS fits a one- or two-factor latent trait model with a logit–probit response function to polytomous data when observations may be missing. These programs being DOS-based (though of course they can be run from within a window), do not have a lot of bells and whistles as some of the others do, but they are extremely easy to use.

9.A.9 lca.S and lcreg.sas

The S-plus function lca (contained in the file lca.S), performs the classical latent class analysis on dichotomous data coded 0–1. The SAS macro lcr (contained in the file lcreg.sas) performs latent class regression (i.e., a MIMIC model). The two files are downloadable from the site http://iosun01.biostat.jhsph.edu/~bjohnson/software/lvar/lvar.html. However, the SAS macro is also limited to response data that are dichotomous and coded 0–1, though covariates can be metric.

9.B COMPUTER PROGRAM LISTINGS FOR THE EXAMPLES

9.B.1 A LEM Program for Comparing Sampling Zeros

```
* Comparing Sampling Zeros among Multiple Groups
* Data: Liao (1989b, Table 6.1)
*
* G is the group variable, relig16
* A, B and C are indicators of attend, pray, and reliten
* A design matrix is used to test equality of the zero cells
```

```
man 4
dim 2 2 3 3
lab G C B A
mod {cov(GABC,1)}

des [0 0 0 0 0 0 0 0 0
     1 1 0 1 0 0 0 1 0
     0 0 0 0 0 0 0 0 0
    -1 -1 0 -1 0 0 0 -1 0]

dat [12 6 2 3 6 3 2 2 2
      0 0 3 0 2 9 1 0 7

    105 68 21 42 59  34 13 12  6
      5 15 42  5 19 104  3 11 51]
```

9.B.2 A LEM Program for Comparing LCMs Assuming Two-Way Interaction

```
* Multiple group latent class analysis
* Data: Liao (1989b, Table 6.1).
*
* G is the group variable, relig16;
* A, B and C are indicators of attend, pray, and reliten.
* Comparing marginal models: {AX} and {BX} are assumed
* fully interactive.

lat 1
man 3
dim 2 2 3 3
lab X C B A
mod X A|X B|X C|X
dat  [12 6 2 3 6 3 2 2 2
       0 0 3 0 2 9 1 0 7]

* dat[105 68 21 42 59  34 13 12  6
*       5 15 42  5 19 104  3 11 51]
```

The program is run twice: once as is, and then with the data for the second group only, the first group data being commented out.

9.B.3 A LEM Program for Comparing LCMs Assuming Linear-by-Linear Association

```
* Multiple group latent class analysis
* Data: Liao (1989b, Table 6.1).
*
* G is the group variable, relig16;
```

COMPUTER PROGRAM LISTINGS FOR THE EXAMPLES   165

* A, B and C are indicators of attend, pray, and reliten.
* Comparing marginal models: {AX} and {BX} association are
* assumed linear-by-linear.

```
lat 1
man 3
dim 2 2 3 3
lab X C B A
mod X A|X {A,spe(AX,1b)} B|X {B,spe(BX,1b)} C|X
dat [12 6 2 3 6 3 2 2 2
     0 0 3 0 2 9 1 0 7]

*dat[105 68 21 42 59 34 13 12  6
*      5 15 42  5 19 104  3 11 51]
```

9.B.4 A LEM Program for Comparing LCMs Assuming Column-Effect RC-II Association

```
*    Multiple group latent class analysis
*    Data: Liao (1989b, Table 6.1).
*
*    G is the group variable, relig16;
*    A, B and C are indicators of attend, pray, and reliten.
*    Comparing marginal models: {AX} and {BX} association are
*    assumed RC-II.

lat 1
man 3
dim 2 2 3 3
lab X C B A
mod X A|X {A,ass2(A,X,4a)} B|X {B,ass2(B,X,4a)} C|X
dat [12 6 2 3 6 3 2 2 2
     0 0 3 0 2 9 1 0 7]

*dat [105 68 21 42 59  34 13 12  6
*       5 15 42  5 19 104  3 11 51]
```

9.B.5 A LEM Program for Comparing LCMs Assuming Complete Homogeneity

```
*    Multiple group latent class analysis.
*    data: Liao (1989b, Table 6.1).
*
*    G is the group variable, relig16;
*    A, B and C are indicators of attend, pray, and reliten.
```

*   Completely homogeneous model: No variable depends on G.

```
lat 1
man 4
dim 2 2 2 3 3
lab X G C B A
mod X A|X B|X C|X
dat [12 6 2 3 6 3 2 2 2
      0 0 3 0 2 9 1 0 7

     105 68 21 42 59  34 13 12  6
       5 15 42  5 19 104  3 11 51]
```

9.B.6 A LEM Program for Comparing LCMs Assuming Complete Heterogeneity

```
*    Multiple group latent class analysis.
*    data: Liao (1989b, Table 6.1).
*
*    G is the group variable, relig16;
*    A, B and C are indicators of attend, pray, and reliten.
*    Completely heterogeneous model: X, A|X, B|X and C|X all
*    depend on G

lat 1
man 4
dim 2 2 2 3 3
lab X G C B A
mod G X|G A|XG B|XG C|XG
dat [12 6 2 3 6 3 2 2 2
      0 0 3 0 2 9 1 0 7

     105 68 21 42 59  34 13 12  6
       5 15 42  5 19 104  3 11 51]
```

9.B.7 A LEM Program for Comparing LCMs Assuming Partial Heterogeneity

```
*    Multiple group latent class analysis.
*    data: Liao (1989b, Table 6.1).
*
*    G is the group variable, relig16;
*    A, B and C are indicators of attend, pray, and reliten.
*    Partially homogeneous model: X as well as A, B, and C
*    depend on G, but AX, BX, and CX do not.
```

COMPUTER PROGRAM LISTINGS FOR THE EXAMPLES 167

```
lat 1
man 4
dim 2 2 2 3 3
lab X G C B A

mod G
   X|G
   A|XG {AX,AG}
   B|XG {BX,BG}
   C|XG {CX,CG}

dat [12  6  2  3  6  3  2  2  2
      0  0  3  0  2  9  1  0  7

    105 68 21 42 59  34 13 12  6
      5 15 42  5 19 104  3 11 51]
```

9.B.8 A LEM Program for Comparing LCMs Assuming Partial Homogeneity

```
*    Multiple group latent class analysis.
*    data: Liao (1989b, Table 6.1).
*
*    G is the group variable, relig16;
*    A, B and C are indicators of attend, pray, and reliten.
*    Partially homogeneous model: X depends on G,
*    but AX, BX, and CX do not.

lat 1
man 4
dim 2 2 2 3 3
lab X G C B A

mod G X|G A|X B|X C|X
dat [12  6  2  3  6  3  2  2  2
      0  0  3  0  2  9  1  0  7

    105 68 21 42 59  34 13 12  6
      5 15 42  5 19 104  3 11 51]
```

9.B.9 Data for Example 2

The data were originally used as a LEM example, which is based on Hagenaars (1993, Table 5.1). The data are frequencies resulting from the cross-classification of the following eight variables:

A = system responsiveness (1: low; 2: high)
B = ideological level (1: ideologues; 2: nonideologues)

C = repression potential (1: low; 2: high)
D = protest approval (1: low; 2: high)
E = conventional participation (1: low; 2: high)
S = sex (1: men; 2: women)
T = education (1: some college; 2: less than college)
G = age (1: 16–34; 2: 35–57; 3: 58–91)

The frequency data are stored in an ASCII file named `hagen93.fre` and called up from the LEM programs given later. The levels of the variable at the rightmost column change the fastest, and those of the variable at the leftmost column change the slowest.

```
hagen93.fre:

 0  0  0  0  0  1  0  0  1  0  1  0
 0  1  0  0  1  0  0  0  0  0  1  0
 4  0  0  1  0  1  1  0  0  2  1  0
 6  3  1  3  1  1  4  4  1  1  1  0
 0  0  1  0  2  1  0  0  0  0  0  0
 2  3  0  0  2  2  0  3  2  0  4  1
 0  1  1  0  1  1  2  1  0  0  0  0
 4  7  4  0  4  2  4  2  0  0  3  2
 0  1  0  8  1  2  1  0  0  5  3  7
 0  1  0  3  5  3  0  0  1  1  2  2
 2  0  1  9  4  2  8  3  0 12  6  1
10  3  1  6  2  1  7  3  1 11  6  3
 0  1  0  2 10 13  3  2  2  8 21 47
 0  3  1  1 11 10  2  1  5  4 13 17
 1  1  2  9  7  7  3  3  0  7 11  8
 1  4  0  3  7  5  1  9  1  3  6  4
 0  0  0  1  0  1  0  0  0  1  0  0
 1  3  2  0  0  2  0  3  0  0  0  1
 2  1  0  0  0  0  2  1  0  1  0  1
11  6  5  2  4  0 13  7  4  1  1  1
 1  1  0  0  0  1  1  0  0  0  0  3
 1  5 10  1  0  5  0  2  4  1  4  5
 3  2  0  0  1  0  2  0  2  0  0  0
10 17  4  0  3  1  5  7  4  3  5  4
 0  0  0  3  0  1  0  2  0  4  2  4
 1  3  0  0  2  1  1  1  3  2  0  2
 2  1  0  7  0  3  6  0  1  8  4  1
13  7  1 10  4  3  9  8  6  4  6  9
 0  1  0  3  4  6  4  0  6  6  4 15
 5  7  4  2  7  8  2  6  9  3 15 24
 6  1  1  3  4  3  3  4  4  7  5  5
14  8  6  6  9  5  7  6  8  6 17  4
```

9.B.10 A LEM Program for Example 2 with Heterogeneous Groups

```
*     Loglinear Path Model with Latent Variables,
*     with All Parameters Different between the Groups
*     Modified Model 4 of Table 5.2 using Sex as Group
*     Data: Hagenaars (1993, Table 5.1)
*     A = System responsiveness
*     B = Ideological level
*     C = Repression potential
*     D = Protest approval
*     E = Conventional participation
*     S = Sex
*     T = Education
*     G = Age

lat 2
man 8
dim 2 2 2 2 2 2 2 3
lab Y Z A B C D E S T G
mod S
    GT|S      {SGT}
    Y|SGT     {STY,SGY}
    Z|SGTY    {STZ,SGZ,SYZ}
    A|SY
    B|SY
    C|SZ
    D|SZ
    E|SY
dat hagen93.fre
```

9.B.11 A LEM Program for Example 2 with Homogeneous Measurement Parameters

```
*     Loglinear Path Model with Latent Variables,
*     with Equal Measurement Parameters between the Groups
*     Modified Model 4 of Table 5.2 using Sex as Group
*     Data: Hagenaars (1993, Table 5.1)
*     A = System responsiveness
*     B = Ideological level
*     C = Repression potential
*     D = Protest approval
*     E = Conventional participation
*     S = Sex
*     T = Education
*     G = Age
```

```
lat 2
man 8
dim 2 2 2 2 2 2 2 2 3
lab Y Z A B C D E S T G
mod S
    GT|S    {SGT}
    Y|SGT   {STY,SGY}
    Z|SGTY  {STZ,SGZ,SYZ}
    A|Y
    B|Y
    C|Z
    D|Z
    E|Y
dat hagen93.fre
```

9.B.12   A LEM Program for Example 2 with Homogeneous Structural Parameters

```
*   Loglinear Path Model with Latent Variables,
*   with Equal Structural Parameters between the Groups
*   Modified Model 4 of Table 5.2 using Sex as Group
*   Data: Hagenaars (1993, Table 5.1)
*   A = System responsiveness
*   B = Ideological level
*   C = Repression potential
*   D = Protest approval
*   E = Conventional participation
*   S = Sex
*   T = Education
*   G = Age

lat 2
man 8
dim 2 2 2 2 2 2 2 2 3
lab Y Z A B C D E S T G
mod S
    GT|S    {SGT}
    Y|GT    {TY,GY}
    Z|GTY   {TZ,GZ,YZ}
    A|SY
    B|SY
    C|SZ
    D|SZ
    E|SY
dat hagen93.fre
```

9.B.13 A LEM Program for Example 2 with Homogeneous Model Parameters

```
*    Loglinear Path Model with Latent Variables,
*    with Equal Constraints for All Model Parameters
*    and Latent Distributions between the Groups
*    Modified Model 4 of Table 5.2 using Sex as Group
*    Data: Hagenaars (1993, Table 5.1)
*    A = System responsiveness
*    B = Ideological level
*    C = Repression potential
*    D = Protest approval
*    E = Conventional participation
*    S = Sex
*    T = Education
*    G = Age

lat 2
man 8
dim 2 2 2 2 2 2 2 3
lab Y Z A B C D E S T G
mod
     GT    {GT}
     Y|GT  {TY,GY}
     Z|GTY {TZ,GZ,YZ}
     A|Y
     B|Y
     C|Z
     D|Z
     E|Y
dat hagen93.fre
```

CHAPTER 10

Comparison in Multilevel Analysis

10.1 INTRODUCTION

In Chapters 3 through 9 we have considered statistical comparison in which groups are external. They are regarded as categories of a fixed variable across which a specific model—a GLM or a model with a latent variable, for example—is compared. Sometimes the entire model is not constrained to be equal across the groups. Instead, a submodel such as the measurement or the structural model is hypothesized to be identical. Thus, this model identity or equality may mean several things. Although often model parameters are of primary interest, researchers may also be interested in hypotheses testing equality of distributional assumptions and model structures.

In this chapter we consider another possibility. We may regard the grouping variable as an internal part of the statistical model instead of as an entity external of it. The distinction is chiefly conceptual rather than practical; for in earlier chapters we have already used interaction terms involving the grouping variable to test equality hypotheses, as one way to set up a fixed effects model. Conceptually, groups can be viewed as units in which individual observations are sampled and measurements made. As such, groups form a *level* in a multilevel analysis.

There are two further possibilities for groups to be included in a multilevel analysis. In cross-national research, individuals are sampled within a number of nations, each of which has distinctive known properties. In such a case, group effects and differentials are better modeled as fixed. This scenario is what we have treated so far. Groups can also be drawn randomly from a larger population, whose members may or may not possess properties known to be distinctive cross the units. This requires a model with random effects.

In this chapter we introduce random effects models for hierarchical data, which are among the major extensions of the GLM (Nelder 1998). These extensions, known as generalized linear mixed models, can be computationally quite demanding, and require special care even when the more advanced adaptive Gaussian quadrature is used, as in SAS's Proc NLMIXED (Lesaffre and Spiessens 2001). Here we do not consider the complex issues of computation and estimation, but focus on the model comparison only, assuming the data analyst has access to an advanced computer program for estimation and follows the instructions for the program. In this context, we examine both fixed and random effects models in the context of statistical group comparison.

The flexibility of multilevel analysis allows the data analyst to model one type of group within another, hence creating models involving three or more levels. Given that the groups are fixed at the higher level of, say, a two-level analysis whose lower level has random effects, a researcher may once again consider the higher-level group as external and compare a two-level model across the units of the higher level. The approach should be equivalent to a three-level mixed model with both fixed and random effects. We discuss this treatment in this chapter too.

10.2 AN INTRODUCTION TO MULTILEVEL ANALYSIS

10.2.1 The Multilevel Setting

The very nature of nested observational phenomena has stimulated interest in and development of multilevel methods for hierarchical modeling in the social and behavioral sciences. In sociology, contextual analysis developed to deal with the same analytic problem faced by researchers using multilevel methods. The method developed in the biomedical sciences is known as mixed-effects models. In economics, the pooling of cross-sectional and time-series data and their analysis also relate to the topic. The wide availability of software for multilevel modeling has in recent years made multilevel analysis easier.

Often the data we collect have a hierarchical or clustered structure. For example, children from the same family tend to be more alike in their physical, mental, and behavioral characteristics than persons chosen at random from the population at large. We may make the same observation about students in the same class or school. The first example in the chapter, analyzing some primary school data, is from the sociology of education, but the techniques and principles are applicable more generally.

We view the hierarchy of data as consisting of units grouped at different levels. Thus children or students may be the level-1 units in a two-level structure where the level-2 units are the families or classes: Children or students may be the level-1 units clustered within families or classes that are the level-2 units. The hierarchy does not have to stop at level 2. We may further consider families as clustered within neighborhoods or communities and classes within schools. Neighborhoods and schools thus constitute level 3 of the hierarchy. Although we focus below on multilevel models with two levels, the generalization to those with three levels is straightforward.

Multilevel data structure is not confined to observational settings. Designed experiments can also create data hierarchies, as in clinical trials that are carried out in a number of randomly chosen centers or groups of individuals. Thus, data analysts apply multilevel models that take into account such hierarchies in order to assess the true treatment effect. In such a clinical-trials study, the subjects are the level-1 units, which are embedded in the level-2 units of the centers chosen for the study.

Oftentimes these level-2 (or higher-level) groups are units within which individuals share many characteristics, such as families, selective schools, and nations. These units can be either fixed (such as nations) or drawn from a random sample (such as families and schools). Other times, however, groupings may arise from individuals who do not share a significant set of characteristics, at least initially—for example, the assignment of young children to public elementary schools or patients to clinics. Nonetheless, membership in these groups over time may forge certain shared characteristics that were not there before. Multilevel analysis is appropriate for both situations.

Traditionally, researchers would analyze multilevel data either at the individual level (by ignoring the level-2 variations), or at the macro level only (by aggregating the data). In either case, serious problems may occur. Aggregate data analysis may lead to the so-called ecological fallacy. Ignoring clustering among individuals contradicts one of the basic assumptions of statistical models such as the generalized linear model: that a random error distribution is i.i.d. Neither approach is desirable.

10.2.2 An Introductory Bibliography

The statistical method known in different fields as multilevel analysis, hierarchical linear models, and mixed-effects models arose for dealing with a multilevel data structure by taking into account its nonindependent errors. There exist many general treatments of the topic, among which at least six are of note:

1. Bryk, A. S. and S. W. Raudenbush. 1992. *Hierarchical Linear Models*: *Applications and Data Analysis Methods*. Newbury Park, CA: Sage.
2. Goldstein, H. 1995. *Multilevel Statistical Models*. London: Edward Arnold. Available in electronic form at
   `http://www.arnoldpublishers.com/support/goldstein.htm`

3. Hox, J. J. 1994. *Applied Multilevel Analysis*. Amsterdam: TT-Publikaties. Available in electronic form at http://www.ioe.ac.uk/multilevel/amaboek.pdf
4. Kreft, I. G. and J. de Leeuw. 1998. *Introducing Multilevel Modeling*. Thousand Oaks, CA: Sage.
5. Longford, N. T. 1993. *Random Coefficient Models*. New York: Oxford University Press.
6. Snijders, T. and R. Bosker. 1999. *Multilevel Analysis: An Introduction to Basic and Advanced Multilevel Modeling*. Thousand Oaks, CA: Sage.

For readers who are interested in an introduction to the topic, the items 3 and 4 are good candidates. Items 2 and 5, on the other hand, include more more mathematical background and details, and item 1 is an extensive and thorough treatment of the hierarchical linear model. The final item on the list offers sufficient coverage of both introductory and updated information on more advanced topics such as hierarchical logit models.

10.2.3 Fixed versus Random Effects

The intercepts, coefficients, or more generally parameters in the random intercept model and the random coefficient model discussed in later sections can be treated as either fixed or random effects that are group-specific. There are no hard and fast rules regarding whether the fixed or the random approach is the appropriate one. However, a couple of properties of the data structure can be considered.

One thing to consider in choosing a fixed versus a random effects model is the number of level-2 units. Snijders and Bosker (1999) discussed a rule of thumb that depends on J, the number of groups in the data. If J is small (say $J < 10$), then use the fixed effects approach; if the groups are viewed as a sample from a population, then the data contain only scanty information about this population. If J is not small ($J \geq 10$) while each group size is small or intermediate (say $n_j < 100$), then use the random effects approach, because 10 or more groups are usually too many to be treated as unique or fixed entities. If the group sizes are large ($n_j \geq 100$), then which view to take makes little difference. Snijders and Bosker (1999) suggested that this rule of thumb should only serve as a first hunch, not as a determining factor for the choice of fixed and random effects models.

In choosing fixed versus random effects models, Longford (1993) recommended an examination of the characteristics of level-2 units in terms of their *exchangeability*. For example, different schools and hospitals are exchangeable in that the inferences we wish to draw do not depend on their identification. Along the same line of argument, the level-1 units embedded within schools and hospitals are exchangeable as well, in that the pupils and patients are exchangeable because the behavioral or medical inferences we wish to draw do not depend on their individual identification. This does not mean either the level-1 or the level-2 units are all identical among themselves. Their uniqueness is captured by measurements such as socioeconomic status for the level-1 units, and school or hospital type (e.g., public versus church related) for the level-2 units. In such a situation the level-2 units are preferably treated as random.

Nations, ethnic groups, and the sexes, on the other hand, are not exchangeable; each has its own unique properties. Each group can be defined and identified unambiguously. These units are preferably regarded as fixed.

The criterion of exchangeability and the rule of sample size do often agree. For example, when we compare nations, ethnic groups, and the sexes, most often we do not have a large number of these groups ($J < 10$). The group comparisons we have discussed in the previous chapters all assumed fixed level-2 units.

10.3 THE BASICS OF THE LINEAR MULTILEVEL MODEL

10.3.1 The Basic Data Structure

We have described in the previous section the situations in which the multilevel data structure is found and multilevel models are needed. Consider the two-level structure that describes the clustering pupils within schools and clustering residents within neighborhoods. The structure of such two-level data (one-way layout) can be summarized by Table 10.1.

The notation in this chapter follows that in the multilevel literature (e.g., Bryk and Raudenbush 1992; Goldstein 1995). Thus, we use the subscript j to indicate level-2 units, though j can be conveniently replaced by g to indicate groups to which observations belong, to be consistent with earlier chapters. Notice that each group or level-2 unit need not have the same number of level-1 observations n_j.

Table 10.1 The Structure of Two-Level Data

Level-2 Unit	Level-1 Units			
1	y_{11}	y_{21}	$\cdots$	$y_{n_1 1}$
2	y_{12}	y_{22}	$\cdots$	$y_{n_2 2}$
3	y_{13}	y_{23}	$\cdots$	$y_{n_3 3}$
$\vdots$	$\vdots$	$\vdots$		$\vdots$
J	y_{1J}	y_{2J}	$\cdots$	$y_{n_J J}$

Note: Each row represents a level-2 unit j, within which i describes the order number of level-1 units, which are identified by their subscripts ij. Level-2 unit j contains n_j elementary units, and the number of level-2 units is J.

THE BASICS OF THE LINEAR MULTILEVEL MODEL   177

When the level-2 units are embedded in a higher level unit, a three-level data structure emerges. Consider Table 10.1 as a subtable of a larger table indexed by k with a total of K level-3 units. Following earlier examples, schools are embedded within school districts, and neighborhoods within towns and cities, for example. Again, each subtable need not have the same number of rows (level-2 units) or columns (level-1 units).

10.3.2 Random Intercept Models

10.3.2.1 One-Way ANOVA

First, let us consider a model without any explanatory variables, which is expressed as

$$y_{ij} = \beta_{0j} + e_{ij}, \tag{10.1}$$

where y_{ij} represents an outcome variable of interest, e_{ij} stands for unexplained errors or residuals, and i and j index the level-1 and level-2 units, respectively. As usual, we assume $e_{ij} \sim N(0, \sigma^2)$, and refer to σ^2 as the level-1 variance. This model indeed is just a one-way ANOVA with random effects, and the intercept, β_{0j}, is just the group means.

At level 2 (the group level), each group-specific mean of the dependent variable, β_{0j}, is represented as a function of the grand mean, γ_{00}, plus a random error, u_{0j}:

$$\beta_{0j} = \gamma_{00} + u_{0j}, \tag{10.2}$$

where we define $E(\beta_{0j}) = \gamma_{00}$, $\text{var}(\beta_{0j}) = \tau_{00}$, and assume $u_{0j} \sim N(0, \tau_{00})$. We refer to τ_{00} as the level-2 variance. An important concept in multilevel analysis using τ_{00} is that of intraclass correlation coefficient, given by the formula

$$\rho = \tau_{00}/(\tau_{00} + \sigma^2). \tag{10.3}$$

All multilevel linear models can be broken down into two components—the fixed component and the random component. The fixed component is represented by the parameters and explanatory variables in the model (though the current model has none), and the random component consists of the two variances σ and τ_{00}.

10.3.2.2 One-Way ANCOVA

Next, let us expand the model of (10.1) and (10.2) by including a certain explanatory variable or covariate in the level-1 equation:

$$y_{ij} = \beta_{0j} + \beta_{1j}x_{ij} + e_{ij}. \tag{10.4}$$

Now we have two level-2 equations, one for each β:

$$\beta_{0j} = \gamma_{00} + u_{0j},$$
$$\beta_{1j} = \gamma_{10}.$$
(10.5)

Notice that the effect of x_{ij} is constrained to be the same for each level-2 unit. Seen from the linear regression point of view, the regression lines of the groups are parallel to one another. Thus, (10.4) and (10.5) define an ANCOVA model with random effects.

The model can be extended to include level-2 covariates. Next, we consider such extension by allowing differences between the within-group and between-group regressions. Here we include a level-2 variable in the level-2 equation for the random intercept:

$$\beta_{0j} = \gamma_{00} + \gamma_{01}w_{1j} + u_{0j},$$
(10.6)

where w_{1j} can be considered the group-specific mean estimate of the outcome. The other two equations as specified in (10.4) and (10.5) remain unchanged.

Here the classical ANOVA assumes that the covariate effect, γ_{10}, is identical for each group. Using matrix notation, such a model can be readily extended to one with multiple explanatory variables in both the level-1 and the level-2 equations. Here we use a column vector to represent the level-1 units in each group or level-2 unit j, thus omitting the subscript i in the level-1 equation, and we use a column vector to represent the level-2 units j for each β from the level-1 equation, thus omitting the subscript j in the level-2 equation:

$$\mathbf{y}_j = \mathbf{X}_j \boldsymbol{\beta}_j + \mathbf{e}_j,$$
$$\boldsymbol{\beta}_k = \mathbf{W}_0 \boldsymbol{\gamma}_k + \mathbf{u}_0,$$
(10.7)

where $\boldsymbol{\beta}_k$ is a vector of J random coefficients, one for each β_k; $\boldsymbol{\gamma}_k$ is a group-specific vector of level-2 regression parameters for each β_k with $k = 0, \ldots, K$ for the level-1 parameters, $\mathbf{u}_0$ is a vector of random errors that contains all zeros for $k = 1, \ldots, K$ but is distributed with $(0, \tau_{00})$ when $k = 0$ (thus the notation $\mathbf{u}_0$); and $\mathbf{X}_j$ and $\mathbf{W}_0$ comprise their respective set of explanatory variables. $\mathbf{W}_0$ contains scalar constants equal to 1 when $k = 1, \ldots, K$, but it has free parameters to estimate when $k = 0$.

So far we have only considered random intercept multilevel models. In the next subsection we relax this assumption and consider multilevel models with random coefficients as well.

10.3.3 Random Coefficient Models

The general random intercept model (10.7) assumes that regression slopes in the level-1 equation are identical to one another. A relaxation of this assumption gives rise to the next submodel.

10.3.3.1 Random Coefficient Models without Cross-Level Effects
Now we allow level-1 regression slopes to be group-specific:

$$\begin{aligned} \mathbf{y}_j &= \mathbf{X}_j \boldsymbol{\beta}_j + \mathbf{e}_j, \\ \boldsymbol{\beta}_k &= \mathbf{W}_0 \boldsymbol{\gamma}_k + \mathbf{u}_k, \end{aligned} \quad (10.8)$$

where $\mathbf{u}_k$ is the same vector of random errors as defined earlier except that it is no longer constrained to be zero for $k = 1, \ldots, K$. Every other element is defined the same as before. This model contains no level-2 covariates for $k = 1, \ldots, K$ other than allowing the level-1 coefficients to be group-specific or level-2 specific. In that sense, the model does not have so-called cross-level coefficients, which are for the interaction between level-1 and level-2 covariates.

If the groups are viewed as fixed, this model is similar to the common approach of including interaction effects of the group variables with all x-variables. Multiplying X and the grouping variable, we obtain

$$\mathbf{y}_j = \mathbf{X}_j \boldsymbol{\beta} + u_{0j} + \mathbf{X}_j \boldsymbol{\gamma}_j + \mathbf{e}_j. \quad (10.9)$$

In many applications the researcher simply include a vector of the main effects for the groups (in place of the random component as represented by u_{0j}) and the interaction effects of $\boldsymbol{\gamma}_j$. Thus, this model can be regarded as the fixed-effects version of the random coefficient model.

10.3.3.2 Random Coefficient Models with Cross-Level Effects
We now extend the model in the previous section by including covariates as well as random errors in each level-2 equation. In matrix notation, we obtain

$$\begin{aligned} \mathbf{y}_j &= \mathbf{X}_j \boldsymbol{\beta}_j + \mathbf{e}_j, \\ \boldsymbol{\beta}_k &= \mathbf{W}_k \boldsymbol{\gamma}_k + \mathbf{u}_k, \end{aligned} \quad (10.10)$$

where not only do we allow for random errors for explaining coefficients other than the intercept, but we also allow any set of w-variables for explaining the variation in each β_k across the groups. In other words, these w-variables form cross-level effects with the x-variables. In fact, (10.10) defines the hierarchical linear model in its general form, and earlier models can all be seen as special cases by constraining certain elements in it.

Table 10.2 A Segment of the Language Score Data

	Individual-Level Data		
Pupil	IQ	Language Test	SES
1	3.16590	46	−4.81200
1	2.66590	45	−17.81200
1	−2.33410	33	−12.81200
⋮	⋮	⋮	⋮
2	−0.33406	21	−12.81200
2	−1.33410	27	−12.81200
2	−3.83410	16	−7.81200
⋮	⋮	⋮	⋮
10	−1.33410	28	−12.81200
10	−1.33410	33	−7.81200
10	0.66594	38	−7.81200
⋮	⋮	⋮	⋮

	Group-Level Data		
Class	Mean IQ	Mixed Grade	Size
1	−1.51410	0	5.9
2	−2.83410	1	−14.1
10	−1.33410	1	−13.1
⋮	⋮	⋮	⋮

Source: http://stat.gamma.rug.nl/.

10.3.4 An Example

To fix the idea of analyzing multilevel data using a linear multilevel model, let us look at an example. The data we analyze below are concerned with 11-year-old pupils in elementary schools in the Netherlands (Snijders and Bosker 1999), and may be obtained from Tom Snijders's homepage (http://stat.gamma.rug.nl/). The sample size with complete records is 2,287 from 131 classes. Class sizes in the current data with nonmissing values range from 4 to 35. The nesting structure is pupils within classes.

The dependent variable is the score on a language test. We are interested in how the score depends on the pupil's intelligence and his or her family's socioeconomic status as well as on a number of class-level variables. The data are in two raw data files, one for the individual level (level 1) and the other, the class level (level 2). Table 10.2 presents a portion of the two data files.

The full data files are available from the source line of Table 10.2 or at http://www.staff.uiuc.edu/~f-liao/multilevel/asaeg11.dat and http://www.staff.uiuc.edu/~f-liao/multilevel/asaeg12.dat. Typically, the data files are rectangular with level-2 ID listed as the first variable. A series multilevel models corresponding to those given in the Sections 10.3.2 and 10.3.3 are estimated using HLM 5.

10.3.5 ANOVA with Random Effects

The first model we estimated is

$$\text{LANGSC}_{ij} = \beta_{0j} + e_{ij},$$
$$\beta_{0j} = \gamma_{00} + u_{0j},$$

where LANGSC is the score on a language test for the pupils. As we saw earlier, multilevel models can be broken down into the two components of the fixed and the random components, and the results are presented in Table 10.3. For the our first model, the empty model, the fixed component has only one parameter, γ_{00}, estimated to be 40.36. The random component is estimated by two variance components: $\sigma^2 = 64.56$ and $\tau_{00} = 19.63$. From them we obtain a ρ-estimate of $19.63/(19.63 + 64.56) = 0.23$. This is fairly high: educational research commonly yields values of ρ ranging from 0.05 to 0.20. There exist two interpretations of ρ. From a level-1 point of view, the estimate shows the correlation among pupils within classes. Alternatively, the estimate measures the proportion of variance in the outcome between level-2 units.

10.3.6 ANCOVA with Random Effects

Now we expand the previous ANOVA model by including a pupil-specific explanatory variable, verbal IQ, in the level-1 equation:

$$\text{LANGSC}_{ij} = \beta_{0j} + \beta_{1j}\text{IQ}_{ij} + e_{ij},$$
$$\beta_{0j} = \gamma_{00} + u_{0j},$$
$$\beta_{1j} = \gamma_{10}.$$

Table 10.3 Estimates from the ANOVA Model with Random Effects

Fixed Effect	Estimate	S.e.
γ_{00}	40.362	0.427
Random Effect	Variance Component	S.d.
τ_{00}	19.633	4.431
σ^2	64.564	8.035
Deviance	16,253.081	

Table 10.4 Estimates from the ANCOVA Model with Random Effects

Fixed Effect	Estimate	S.e.
γ_{00}	40.608	0.305
γ_{10}	2.488	0.081
Random Effect	Variance Component	S.d.
τ_{00}	9.601	3.099
σ^2	42.245	6.500
Deviance	15,253.934	

Here we have two level-2 equations, one for each β in the level-1 equation. So far we have only considered random intercept multilevel models. That is, we have only allowed the intercept to vary randomly but kept the coefficient of the explanatory variable fixed to be equal among the groups. The parameter estimate for IQ (γ_{10}) is 2.49 with a standard error of 0.08, a highly significant estimate. The intraclass correlation coefficient based on the results in Table 10.4 decreases to 0.19.

The identical slopes for all groups, together with the varying intercepts from the first level-2 equation, suggest that the regression lines are all parallel to one another but may cross the Y-axis at different locations.

We now further extend the model by including a covariate in the level-2 equation for the random intercept:

$$\text{LANGSC}_{ij} = \beta_{0j} + \beta_{1j}\text{IQ}_{ij} + e_{ij},$$

$$\beta_{0j} = \gamma_{00} + \gamma_{01}\text{GIQ}_j + u_{0j},$$

$$\beta_{1j} = \gamma_{10},$$

where GIQ_j is the group-specific mean IQ for each class. It is obvious that the class-specific mean of IQ explains a significant amount of variation in the intercept, with an estimate of 1.59 (γ_{01}) and a standard error of 0.32 (Table 10.5). The ρ-estimate decreases further to 0.16. The within-group regression coefficient γ_{10} is 2.41, and the between-group regression coefficient is the sum of the within-group regression coefficient γ_{10} and the regression coefficient γ_{01} of the group mean, or 2.41 + 1.59 = 4.00. A pupil with a given level of IQ on average receives a higher language test score (by 1.59) in a class with a higher average IQ. Here the regression slopes of IQ still are parallel to one another among the classes, but the between-group regression slope is much steeper. The reverse can also be true in another research setting. That is, the between-group regression line is less steep than the within-group regression lines.

THE BASICS OF THE LINEAR MULTILEVEL MODEL

Table 10.5   Estimates from the Extended ANCOVA Model with Random Effects

Fixed Effect	Estimate	S.e.
γ_{00}	40.741	0.284
γ_{01}	1.589	0.320
γ_{10}	2.415	0.085
Random Effect	Variance Component	S.d.
τ_{00}	7.886	2.808
σ^2	42.172	6.494
Deviance	15,232.190	

10.3.7   Random Coefficient Models without Cross-Level Effects

We next extend the previous model by including group-level randomness into the level-2 equation for β_{1j}:

$$\text{LANGSC}_{ij} = \beta_{0j} + \beta_{1j}\text{IQ}_{ij} + e_{ij},$$
$$\beta_{0j} = \gamma_{00} + \gamma_{01}\text{GIQ}_j + u_{0j},$$
$$\beta_{1j} = \gamma_{10} + u_{1j}.$$

If the groups can be viewed as fixed, this model in effect is the fixed-effects random coefficient model presented in an earlier section, and it is tantamount to estimating a GLM with interactions between the grouping variable and the x-variables in Chapter 6. An estimation of the model (Table 10.6) gives the average effect of IQ on language test score (γ_{10}) as 2.46, and the standard deviation of this slope, computed as $\sqrt{\text{var}(u_{1j})}$, is 0.46. The value of the average effect (± 2 standard deviations) ranges from 1.54 to 3.38. This suggests that, although IQ always has a positive effect on language score, large effects are more than twice as large as small effects. In this case, the between-group regression line is steeper than some within-group regression lines but less so than the others, since now the within-group regression slopes can be different from one another.

10.3.8   Random Coefficient Models with Cross-Level Effects

We further extend the model by including a level-2 variable, the class size GSIZE, into both level-2 equations:

$$\text{LANGSC}_{ij} = \beta_{0j} + \beta_{1j}\text{IQ}_{ij} + e_{ij},$$
$$\beta_{0j} = \gamma_{00} + \gamma_{01}\text{GIQ}_j + \gamma_{02}\text{GSIZE}_j + u_{0j},$$
$$\beta_{1j} = \gamma_{10} + \gamma_{11}\text{GSIZE}_j + u_{1j}.$$

Table 10.6 Estimates from a Random Coefficient Model without Cross-Level Effects

Fixed Effect	Estimate	S.e.
γ_{00}	40.750	0.286
γ_{01}	1.412	0.345
γ_{10}	2.460	0.086
Random Effect	Variance Component	S.d.
τ_{00}	8.077	2.842
τ_{11}	0.212	0.460
σ^2	41.339	6.430
Deviance	15,217.874	

Table 10.7 Estimates from a Random Coefficient Model with Cross-Level Effects

Fixed Effect	Estimate	S.e.
γ_{00}	40.923	0.289
γ_{01}	1.224	0.350
γ_{02}	0.072	0.039
γ_{10}	2.444	0.084
γ_{11}	-0.025	0.012
Random Effect	Variance Component	S.d.
τ_{00}	7.919	2.814
τ_{11}	0.199	0.446
σ^2	42.310	6.427
Deviance	15,222.972	

The parameter γ_{11} captures the so-called cross-level interaction between class size and individual IQ. Class size ranges from 5 to 37. The negative estimate suggests greater effects of IQ on language test score in smaller classes (Table 10.7).

The model above is an example of the general form of linear multilevel models, which may include as many variables as one wishes to include as long as the model is identified. For the current example we may include all the variables available in the data. This, however, may be neither statistically efficient nor substantively sensible to do. One common approach is to include all reasonable variables first, then trim the model down to a more parsimonious one, using a criterion such as the LRT in GLMs. We present below one such trimmed model:

$$\text{LANGSC}_{ij} = \beta_{0j} + \beta_{1j}\text{IQ}_{ij} + \beta_{2j}\text{SES}_{ij} + e_{ij},$$

$$\beta_{0j} = \gamma_{00} + \gamma_{01}\text{GIQ}_j + \gamma_{02}\text{MGRADE}_j + u_{0j},$$

$$\beta_{1j} = \gamma_{10} + \gamma_{11}\text{MGRADE}_j + u_{1j},$$

$$\beta_{2j} = \gamma_{20},$$

where a new level-1 variable SES and a new level-2 variable MGRADE, representing the status of whether a class has mixed grades, are included in the model, whereas the level-2 variable GSIZE is omitted.

Using full maximum likelihood estimation (to facilitate a LRT) instead restricted likelihood estimation as for the earlier examples, this model yields a deviance score of 15,086.33 (Table 10.8). A model with GIQ, GSIZE, and MGRADE plus a random error included in all the three level-2 equations yields a full maximum likelihood deviance score of 15,079.58. The trimmed model does not decrease the fit to the data significantly, yielding a LRT of 15,086.33 − 15,079.58 = 6.75 with 19 − 10 = 9 degrees of freedom. We have saved nine degrees of freedom but have lost little in explaining the variation in the outcome. The estimates involving the variable of mixed grade suggest that while pupils in mixed-grade classes tend to have lower average language test scores, the effect of IQ tends to be higher for these mixed-grade classes. Here we have a significant cross-level effect between the mixed grade status of a class and individual IQ.

Table 10.8 ML Estimates from a Trimmed Random Coefficient Model with Cross-Level Effects

Fixed Effect	Estimate	S.e.
γ_{00}	41.453	0.330
γ_{01}	0.833	0.328
γ_{02}	−1.714	0.614
γ_{10}	2.079	0.090
γ_{11}	0.542	0.179
γ_{20}	0.156	0.014
Random Effect	Variance Component	S.d.
τ_{00}	7.518	2.742
τ_{11}	0.118	0.343
σ^2	39.281	6.267
Deviance	15,086.328	

10.3.9 Assumptions of the Linear Multilevel Model

The basic assumptions of the linear multilevel model extend those of the linear regression model. They include:

Specification. Do the fixed and the random components of the model together contain all necessary variables?

Distribution of Errors. Do the random errors at all levels follow the normal distribution individually (with level-1 random errors distributed within the units of higher levels)?

Homogeneity of Variances. Do the random errors at all levels have a constant covariance with respect to their covariates?

Independence of the Variables and the Residuals. Are the variables and the residuals uncorrelated with each other at each level of the analysis?

Cross-Level Independence. Are the random errors at a particular level independent of the random errors at the other levels? By extension, are they also independent of the variables at the other levels?

Positive answers to these questions satisfy the basic assumptions of the linear multilevel model.

10.4 THE BASICS OF THE GENERALIZED LINEAR MULTILEVEL MODEL

Often we have dependent variables that are discrete in applications, such as dichotomous, nominal, ordinal, or event count. Using the GLM we discussed in Chapters 6 and 7, the model discussed in the previous sections can be generalized into the class of hierarchical generalized linear models or generalized linear multilevel models (GLMMs), with which such noncontinuous dependent variables can be studied. The estimation of the GLMM has been worked out and well discussed (e.g., McCullogh 1997; Schall 1991). Many multilevel software packages also implement estimations of the GLMM. The GLMM can be extended to include latent variables, and form a class of models known as generalized linear latent and mixed models (GLLAMMs). These can model multivariate responses of mixed type including continuous, dichotomous, ordered and unordered categorical responses, counts, survival data, and rankings.

Following our notation in earlier chapters, the level-1 equation of the GLMM for a two-level situation can be written as

$$y_j = \mu_j + e_j, \tag{10.11}$$

THE BASICS PF THE GENERALIZED LINEAR MULTILEVEL MODEL   187

where j indexes the level-2 units and other symbols are defined as in Chapter 6. Let the $g(\cdot)$ be the monotonic link function, as before, such that

$$g(\boldsymbol{\mu}_j) = \boldsymbol{\eta}_j = \mathbf{X}_j \boldsymbol{\beta}_j. \tag{10.12}$$

However, for the vector of $\boldsymbol{\beta}$ we further define

$$\boldsymbol{\beta}_j = \mathbf{W}_j \boldsymbol{\gamma}_j + \mathbf{u}_j. \tag{10.13}$$

For example, when the level-1 data are binomial and the link function is the canonical logit, for a two-level data structure we obtain

$$\mathbf{y}_j = \mathbf{P}_j + \mathbf{e}_j, \tag{10.14}$$

$$\ln \frac{\mathbf{P}_j}{1 - \mathbf{P}_j} = \mathbf{X}_j \boldsymbol{\beta}_j, \tag{10.15}$$

$$\boldsymbol{\beta}_j = \mathbf{W}_j \boldsymbol{\gamma}_j + \mathbf{u}_j, \tag{10.16}$$

where $\mathbf{P}_j = P(\mathbf{y}_j = 1)$. Other level-1 data distributions and corresponding link functions can be written out without complication.

10.4.1 A Random Coefficient Logit Model with Cross-Level Effects

We revisit the Dutch school example earlier, but recode the language test score variable into a dummy variable, coded 1 if a pupil scored 30 or above.

Table 10.9 HLM Estimates from a Random Coefficient Logit Model with Cross-Level Effects

Fixed Effect	Estimate	S.e.
γ_{00}	2.518	0.097
γ_{01}	0.428	0.125
γ_{10}	0.551	0.040
γ_{11}	0.145	0.082
Random Effect	Variance Component	S.d.
τ_{00}	0.459	0.677
τ_{11}	0.007	0.085
Deviance	996.957	

Table 10.10 SAS Estimates from a Random Coefficient Logit Model with Cross-Level Effects

Fixed Effect	Estimate	S.e.
γ_{00}	2.775	0.142
γ_{01}	0.494	0.134
γ_{10}	0.588	0.057
γ_{11}	0.185	0.098
Random Effect	Variance Component	S.e.
τ_{00}	0.544	0.232
τ_{11}	0.013	0.027
τ_{01}	−0.047	0.063
$-2\mathcal{L}$	1,274.5	

The model fitted in Table 10.9 is a smaller version of that in Table 10.8:

$$\ln\left[\frac{P(\text{LANGPASS}_{ij} = 1)}{1 - P(\text{LANGPASS}_{ij} = 1)}\right] = \beta_{0j} + \beta_{1j}\text{IQ}_{ij},$$

$$\beta_{0j} = \gamma_{00} + \gamma_{01}\text{MGRADE}_j + u_{0j},$$

$$\beta_{1j} = \gamma_{10} + \gamma_{11}\text{MGRADE}_j + u_{1j}.$$

The results are obtained by HLM 5, and those from the population average model with robust standard errors are used. We estimated the same model using SAS PROC NLMIXED. In contrast with HLM, which implements the restricted likelihood and the full likelihood estimation, the SAS procedure relies on likelihood-based estimation using adaptive Gaussian quadrature, and gives more accurate estimates, though the estimation can be more time consuming. The SAS estimates are presented in Table 10.10. The two sets of results, though comparable, are not identical.

We next study a multilevel example in which countries are used as level-2 units. Kluegel, Mason, and Wegener (1995) surveyed 11 European nations plus Japan and the U.S. on social justice issues. One of the questions they asked was about the government's role in personal income restrictions. The respondents were asked their opinions of the statement, "The government should place an upper limit on the amount of money any one person can make." The original five categories are collapsed into 1 = agree and 0 = disagree, since there is a strong tendency toward bipolar distribution when examined at the aggregate level. For the level-1 model we use only two explanatory factors—gender and education (which are represented by three

THE BASICS PF THE GENERALIZED LINEAR MULTILEVEL MODEL

Table 10.11 SAS Estimates from Two GLMMs with Cross-Level Effects

Fixed Effect	Logit		Probit	
	Estimate	S.e.	Estimate	S.e.
β_{2j}	0.551	0.045	0.340	0.0278
β_{3j}	0.864	0.050	0.532	0.030
β_{4j}	0.990	0.049	0.609	0.030
γ_{00}	−1.425	0.319	−0.872	0.195
γ_{01}	5.113	2.212	3.105	1.351
γ_{10}	0.724	0.137	0.446	0.083
γ_{11}	−1.743	0.953	−1.085	0.575
Random Effect	Variance Component	S.e.	Variance Component	S.e.
τ_{00}	0.277	0.112	0.103	0.042
τ_{11}	0.036	0.021	0.013	0.008
τ_{01}	−0.032	0.036	−0.012	0.013
$-2\mathcal{L}$	20,766		20,764	

dummy variables). The level-2 model contains one explanatory variable, unemployment rate in 1991, for each of the two equations:

$$\ln\left[\frac{P(\text{GVTLIMIT}_{ij} = 1)}{1 - P(\text{GVTLIMIT}_{ij} = 1)}\right] = \beta_{0j} + \beta_{1j}\text{GENDER}_{ij} + \beta_{2j}\text{MEDVOC}_{ij}$$

$$+ \beta_{3j}\text{2NDEDUC}_{ij} + \beta_{4j}\text{3RDEDUC}_{ij},$$

$$\beta_{0j} = \gamma_{00} + \gamma_{01}\text{UNEMPLOY}_j + u_{0j},$$

$$\beta_{1j} = \gamma_{10} + \gamma_{11}\text{UNEMPLOY}_j + u_{1j},$$

where GVTLIMIT is the dichotomous dependent variable; GENDER is 1 if male, 0 if female; and the three education variables indicate the respondent has received some form of medium vocational, secondary, or tertiary education, respectively, with primary-school education or below as the reference category. UNEMPLOY, the unemployment rate, is the only level-2 variable. The data for this and following analysis are based 16,378 cases. The results, estimated with SAS PROC NLMIXED, are presented in Table 10.11, and the SAS syntax for a GLMM with a logit link is given in Section 10.B.1.

10.4.2 A Random Coefficient Probit Model with Cross-Level Effects

We also fit the GLMM with a probit link,

$$\Phi^{-1}\left(P(\text{GVTLIMIT}_{ij} = 1)\right) = \beta_{0j} + \beta_{1j}\text{GENDER}_{ij} + \beta_{2j}\text{MEDVOC}_{ij}$$
$$+ \beta_{3j}\text{2NDEDUC}_{ij} + \beta_{4j}\text{3RDEDUC}_{ij},$$
$$\beta_{0j} = \gamma_{00} + \gamma_{01}\text{UNEMPLOY}_j + u_{0j},$$
$$\beta_{1j} = \gamma_{10} + \gamma_{11}\text{UNEMPLOY}_j + u_{1j}.$$

The results are presented in the columns next to those from the logit model in Table 10.11. For the fixed effects parameters, one can easily convert a probit estimate to its logit counterpart by multiplying by a constant somewhere between 1.6 and $\sqrt{3}$, though the same cannot be done for the variances of the random effects.

10.5 GROUP AS AN EXTERNAL VARIABLE IN MULTILEVEL ANALYSIS

It is possible to include another grouping variable to form a higher level of analysis. For example, a cross-national study of social justice in Europe, United States, and Japan may include a level-3 variable, indexed by k, of whether a country was formerly a communist one. Because there are two units in this level—a nation can be either postcommunist or capitalist—an analysis with three levels is senseless and impractical. However, we may treat this variable as an external variable, and conduct group comparison as in previous chapters.

Specifically, the $-2\log$ likelihood from the model based on the individuals from all nations in the study in Section 10.4.1 (see Table 10.11) and those from the two models of the two groups analyzed separately are used to form an LRT (4.15). A significance test at a chosen level of α suggests that the two former ideological groups have effects different enough to warrant individual attention.

This test is general, and not all parameters need be tested simultaneously. For example, fixed effects may be tested for equality while the random components are constrained to be certain identical fixed values for the groups, and vice versa. Subsets of fixed (or random) parameters may be tested for equality across the groups, just as illustrated in Chapters 8 and 9. Finally, the external variable approach can accommodate groups on more than a single level of analysis, as long as they can reasonably be considered as fixed and the total number of them is manageable.

10.6 THE RELATION BETWEEN MULTILEVEL ANALYSIS AND GROUP COMPARISON

10.6.1 Bridging Fixed and Random Effects Models

Let us begin with a simple two-level model with one level-1 and one level-2 explanatory variables:

$$y_{ij} = \beta_{0j} + \beta_{1j} x_{ij} + e_{ij},$$

$$\beta_{0j} = \gamma_{00} + \gamma_{01} w_j + u_{0j},$$

$$\beta_{1j} = \gamma_{10} + \gamma_{11} w_j + u_{1j}.$$

As before, $E(\beta_{0j}) = \gamma_{00}$, $\text{var}(\beta_{0j}) = \tau_{00}$, $E(\beta_{1j}) = \gamma_{10}$, $\text{var}(\beta_{1j}) = \tau_{11}$, and $\text{cov}(\beta_{0j}, \beta_{1j}) = \tau_{01}$. Replacing the level-1 parameters with the level-2 terms, we obtain a combined model

$$y_{ij} = \gamma_{00} + \gamma_{01} w_j + \gamma_{10} x_{ij} + \gamma_{11} w_j x_{ij} + u_{0j} + u_{1j} x_{ij} + e_{ij}, \quad (10.17)$$

where the term u_{0j} gives the random intercept, and the random error u_{1j} is the random coefficient for x_{ij}.

Suppose that the level-2 variable w_j represents nothing but the second-level groups. Because now w_j is categorical, we must change its coefficients to go along with the categories, thereby absorbing the u_{0j} and u_{1j} terms into the coefficients γ_{01} and γ_{11}, if we add subscript j to the coefficients so that the new combined model becomes

$$y_{ij} = \gamma_{00} + \gamma_{01j} w_j + \gamma_{10} x_{ij} + \gamma_{11j} w_j x_{ij} + e_{ij}, \quad (10.18)$$

where j goes from 1 (the first group) to J (the last group). (10.18) gives a fixed-effects model, and is equivalent to the one-model test of group differences involving interactions introduced in Chapters 3 and 6. The parameters γ_{01j} jointly test intercept or mean differences across the J groups, and the parameters γ_{11j} jointly test slope or coefficient differences across the J groups.

When level-2 groups are numerous and there are no unique characteristics distinguishing one group from another as there are in the example of the Dutch school data, the random effects model is obviously preferred. When making group comparisons in research situations where the number of groups is small (such as two to five, as illustrated in previous chapters), the fixed effects model is the clear choice. Sometimes, however, a research problem may fall somewhere between the two ideal scenarios. The cross-national study of social justice examined earlier is such a case. Although nations can qualify for a fixed effects model, the researchers did not include every

European nation or every postcommunist nation in the population in the study. In that sense, some kind of (convenience) sample was drawn. The number of level-2 units (13) also makes a fixed effects model somewhat cumbersome. Judging from the number of level-1 units (greater than 100 for each j) and the number of level-2 units (greater than 10), the choice of fixed and random effects models should make little difference (Snijders and Bosker 1999). In the next section we compare a fixed effects and a random effects model of the social justice data examined earlier in the chapter.

10.6.2 An Example

We first fit a GLMM with logit link to the cross-national social justice data, assuming fixed effects. The model follows the combined model form (10.18), and can be expressed as

$$\ln\left[\frac{P(\text{GVTLIMIT}_{ij} = 1)}{1 - P(\text{GVTLIMIT}_{ij} = 1)}\right]$$
$$= \gamma_{00} + \gamma_{01j}\text{NATION}_j + \gamma_{10}\text{GENDER}_{ij} + \gamma_{20}\text{MEDVOC}_{ij}$$
$$+ \gamma_{30}\text{2NDEDUC}_{ij} + \gamma_{40}\text{3RDEDUC}_{ij} + \gamma_{11j}\text{NATION}_j\text{GENDER}_{ij}$$
$$+ \gamma_{21j}\text{NATION}_j\text{MEDVOC}_{ij} + \gamma_{31j}\text{NATION}_j\text{2NDEDUC}_{ij}$$
$$+ \gamma_{41j}\text{NATION}_j\text{3RDEDUC}_{ij},$$

where the categorical variable NATION is coded as a deviation, so that the main effects γ_{10} to γ_{40} give the mean values of the coefficients. The model is estimated using SPSS. In addition, a GLMM with logit link of the same data assuming random effects is estimated using SAS PROC NLMIXED. The model follows (10.17) without any w-variable and can be expressed as

$$\ln\left[\frac{P(\text{GVTLIMIT}_{ij} = 1)}{1 - P(\text{GVTLIMIT}_{ij} = 1)}\right] = \gamma_{00} + \gamma_{10}\text{GENDER}_{ij} + \gamma_{20}\text{MEDVOC}_{ij}$$
$$+ \gamma_{30}\text{2NDEDUC}_{ij} + \gamma_{40}\text{3RDEDUC}_{ij}$$
$$+ u_{0j} + u_{1j}\text{GENDER}_{ij} + u_{2j}\text{MEDVOC}_{ij}$$
$$+ u_{3j}\text{2NDEDUC}_{ij} + u_{4j}\text{3RDEDUC}_{ij},$$

where the u_{1j} to u_{4j} terms are the random effects in the level-1 coefficients. To simplify the model and the estimation, the covariances of u_0, u_1, u_2, u_3, u_4, u_5 are assumed to be zero. The SAS syntax is given in Section 10.B.3. The results from these two models are presented in Table 10.12. From them it is clear that the fixed effects model and the random effects model yielded

Table 10.12 SAS Estimates from Two GLMMs with Cross-Level Effects

Parameter	Fixed Effects Model		Random Effects Model	
	Estimate	S.e.	Estimate	S.e.
γ_{00}	−0.754	0.040	−0.765	0.179
γ_{10}	0.523	0.036	0.510	0.067
γ_{20}	0.465	0.053	0.516	0.069
γ_{30}	0.852	0.058	0.856	0.082
γ_{40}	1.032	0.056	1.034	0.105
Level-2 Effect	Wald χ^2	df	Variance Component	S.e.
τ_{00}	300.030	12	0.398	0.162
τ_{11}	48.224	12	0.043	0.024
τ_{22}	34.746	12	0.029	0.025
τ_{33}	45.001	12	0.049	0.032
τ_{44}	75.335	12	0.108	0.056
$-2\mathcal{L}$	20,542		20,742	

Note: For the fixed effect model, the Wald χ^2 statistics for the joint significance tests of with respect to the intercept and the level-1 variables are reported.

almost identical parameter estimates, with only the γ_{20} estimate noticeably different between the two models. However, the fixed effects model consistently underestimated the standard errors of the estimates, whether the "average" effects or the level-2 effects. Neither model can be labeled "correct" or "incorrect" in this case. Obviously, the choice of the model lies in the hands of the researcher, who may emphasize the uniqueness in the set of characteristics of each nation and the fixed effects model or instead regard the nations as coming from a larger population and use the random effects model.

The example also illustrates an important point made at the outset of the chapter—statistical comparison involving groups can be carried out as a one-level GLM or as a GLMM. A one-level GLM with coefficients for comparisons across groups is identical to a GLMM with fixed effects. However, in the multilevel framework, effects can also be modeled as random.

10.7 MULTIPLE MEMBERSHIP MODELS

Despite the common practice, sometimes level-1 units cannot be neatly classified into level-2 units. For example, students may have belonged to a combination of schools sequentially (or even nonsequentially), and an individual may belong to more than one household. To deal with this complication, the multiple membership model is used. The rationale of such model is akin to models applying fuzzy set theory (Liao 1989c; Manton, Liu, and Cornelius 1985; Manton et al. 1987). That is, a member may only partially belong in a set.

Consider the example of schools, and suppose that we know the membership of each student in each school. Practically the membership is handled by assigning a weight w_{ij} to each student i, such that for each student $\sum_{j=1}^{J} w_{ij} = 1$. The weights may be assigned proportionally to the length of time a student is in a particular school. When all level-1 units have a weight of 1, we obtain the usual multilevel model. More generally, line 1 of (10.7) is extended to

$$\mathbf{Y}_j = \mathbf{X}_j \boldsymbol{\beta}_j + \mathbf{w}_j \mathbf{v}_j + \mathbf{e}_j, \tag{10.19}$$

where $\mathbf{w}_j$ is a vector of weights of length I, and $\mathbf{v}_j$ is a vector of random errors of length I. For further details, see Hill and Goldstein (1998) and Goldstein (2000).

Because of the explicit treatment of an individual's membership function as noncrisp (i.e., not simply a 0 or 1 function), the multiple membership model has great potential for applications where certain individuals do not have a single identity. Indeed, this line of thinking can apply to other types of model comparison, not just within the multilevel framework.

10.8 SUMMARY

In this chapter we have introduced multilevel models, which may treat groups as a level in the analysis. For the general form of GLMM, we have

$$\mathbf{y}_{jk} = g^{-1}(\boldsymbol{\eta}_{jk}) + \mathbf{e}_{jk},$$

$$\boldsymbol{\eta}_{jk} = \mathbf{X}_{jk} \boldsymbol{\beta}_{jk}, \tag{10.20}$$

$$\boldsymbol{\beta}_{jk} = \mathbf{W}_{jk} \boldsymbol{\gamma}_{jk} + \mathbf{u}_{jk},$$

where k indicates a higher level grouping variable which can be treated as a third level but can be conveniently analyzed as an external grouping variable.

In using GLMM for conducting group comparisons, two important issues must be considered. The data analyst must decide whether to treat a grouping variable as a distinct level of analysis, and whether to use fixed effects or random effects for the group differentials. While there are no hard and fast rules, exchangeability of the units or groups, number of the groups or level-2 units in the analysis, and parsimony of the statistical model can serve as general guidelines.

10.A SOFTWARE FOR MULTILEVEL ANALYSIS

10.A.1 Special-Purpose Software

10.A.1.1 HLM
The most popular software for multilevel analysis, the Windows version of HLM, is easy to use and user-friendly. In version 5, HLM allows two- or three-level analysis, and the dependent variable can be continuous or categorical (dichotomous, ordinal, multinomial, or Poisson). The company producing HLM, Scientific Software International, Inc., has a Web site: http://www.ssicentral.com/hlm/hlm.htm. The student version can be downloaded for free.

10.A.1.2 MIXREG, MIXOR, MIXNO, MIXPREG, MIXGSUR
Compiled by Don Hedeker of the University of Illinois at Chicago, these programs take only ASCII files for data input. MIXREG is for mixed-effects linear regression with autocorrelated errors. MIXOR is for multilevel analysis of dichotomous and ordinal outcomes. MIXOR is for multinomial logistic regression in the multilevel setting, MIXPREG estimates multilevel Poisson regression models, and MIXGSUR analyzes multilevel grouped-time survival data. Visit his Web site for more information or for the software: http://www.uic.edu/~hedeker/mix.html.

10.A.1.3 MLn / MLwiN
Developed by researchers at the London Institute of Education, this is the most extensive multilevel software. Unlike HLM, MLn (or its Windows version, MLwiN), does allow for data manipulation, graphing, and some simple statistical analysis in addition to multilevel analysis. It also allows for categorical dependent variables, the use of macros, and analysis of an arbitrary number of levels. Their Web site is based at the Institute of Education, University of London: http://www.ioe.ac.uk/mlwin/.

10.A.1.4 VARCL
Written by Nicholas Longford, VARCL has two programs, VARCL3 and VARCL9. The latter allows for analysis of up to nine levels with fixed coefficients only, whereas the former allows for analysis up to three levels with random coefficients. The dependent variable can be normally or nonnormally distributed (such as Poisson, binomial, and gamma). For more information, visit: http://www.assess.com/VARCL.html.

10.A.2 General-Purpose Software

10.A.2.1 SAS / Proc Mixed
Unlike other general-purpose software, SAS allows for fitting quite complex multilevel models with all corresponding statistics. A very nice introduction to using Proc Mixed is found in an article by Judith Singer (1998, *Journal of*

Educational and Behavioral Statistics 24: 323–355). Going beyond the linear model, such as modeling discrete outcome variables, is more complicated; the user must run a SAS macro, GLIMMIX or NLINMIX, for such analysis. The macros call Proc Mixed iteratively and make estimates by using restricted or residual pseudo likelihood. A new SAS procedure, Proc NLMixed, also estimates GLMMs by using adaptive Gaussian quadrature. This procedure is more flexible, and allows both the fixed and random components to be nonlinear. Thus it can estimate a variety of standard and user-defined nonlinear mixed (effects) models. The procedure estimates models by directly maximizing an approximate likelihood. http://www.sas.com.

10.A.2.2 BMDP
This program contains several modules for multilevel analysis: BMDP-3V for random intercept models, and BMDP-4V and BMDP-5V for longitudinal analysis, of which the latter allows for random coefficient models. http://www.spssscience.com/Bmdp/.

10.A.2.3 S-plus / NLE and NLME
The NLE and NLME software comprises a set of S (or S-plus) functions for the analysis of linear and nonlinear mixed-effects models, respectively. Both the NLE and NLME software is available for Unix and Windows platforms, and is currently included with the release of S-plus 2000. However, it cannot handle models with categorical response variables. NLME uses both maximum likelihood and restricted maximum likelihood estimation. The data can be grouped according to one or more nested factors. http://www.splus.mathsoft.com/.

You may obtain the NLME software separately from Bell Labs if your system does not already have it: http://cm.bell-labs.com. On this Web site NLME is available for both Unix and Windows systems.

10.A.2.4 Stata
Stata has several modules for multilevel analysis, such as `loneway` for the empty model, `xt` for longitudinal data, `xtreg` for linear random intercept model, and `xtpois`, `xtnbreg`, `xtlogit`, and `xtprobit` for the multilevel Poisson and negative binomial regression and logit and probit regression models. The last two modules are based on the generalized estimating equations (GEE) method. http://www.stata.com/.

In addition, a Stata program that fits GLLAMM models can be obtained from Sophia Rabe-Hesketh's Web site: http://www.iop.kcl.ac.uk/IoP/Departments/BioComp/programs/gllamm.html.

10.A.3 Software for Other Special Purposes

10.A.3.1 LISREL
The latest version, LISREL 8, allows for certain types of multilevel analysis, but you must have the structural equation mindset. Find out more at the Web site: http://www.ssicentral.com/lisrel/mainlis.htm.

10.A.3.2 MPlus
Another program for covariance structure analysis, MPlus allows for two-level data using the hierarchical model as well as the path-analytic model. For more information, visit http://www.statmodel.com/.

10.A.3.3 BUGS
Developed by researchers at University of Cambridge and Imperial College, this software implements the Gibbs sampler. It requires balanced data for multilevel analysis. However, it is flexible for estimating a large variety of statistical models. Visit their Web site:
http://www.mrc-bsu.cam.ac.uk/bugs/welcome.shtml

10.B SAS PROGRAM LISTINGS FOR GLMM EXAMPLES

10.B.1 Syntax for Producing the Logit Results in Table 10.11

```
options ls = 80 ps = 60; data one; infile 'isjall.dat';
   input country gender bigvtlim jp average unemrate westeur
   postcom centeur easteur medvoc seceduc thdeduc;
proc nlmixed;
   parms gamma00 = 2.5 gamma01 = 0.5 gamma10 = 0.5 gamma11 = 0.1
      su02 = 1 su12 = 1 cu12 = 0.5 beta3 = 0.5 beta4 = 0.5;
   beta0 = gamma00 + gamma01*unemrate + u0;
   beta1 = gamma10 + gamma11*unemrate + u1;
   eta = beta0 + beta1*gender + beta2*medvoc + beta3*seceduc +
beta4*thdeduc;
   expeta = exp(eta);
   p = expeta / (1 + expeta);
   model bigvtlim ~ binary(p);
   random u0 u1 ~ normal([0,0],[su02,cu12,su12]) subject =
country;
 run;
```

10.B.2 Syntax for Producing the Probit Results in Table 10.11

```
options ls = 80 ps = 60; data one; infile 'isjall.dat';
   input country gender bigvtlim jp average unemrate westeur
   postcom centeur easteur medvoc seceduc thdeduc;
proc nlmixed;
   parms gamma00 = 2.5 gamma01 = 0.5 gamma10 = 0.5 gamma11 = 0.1
      su02 = 1 su12 = 1 cu12 = 0.5 beta3 = 0.5 beta4 = 0.5;
   beta0 = gamma00 + gamma01^nemrate + u0;
   beta1 = gamma10 + gamma11^nemrate + u1;
```

```
  eta = beta0 + beta1*gender + beta2*medvoc + beta3*seceduc +
  beta4*thdeduc;
  p = probnorm(eta);
  model bigvtlim ~ binary(p);
  random u0 u1 ~ normal([0,0],[su02,cu12,su12]) subject =
country;
run;
```

10.B.3 Syntax for the Random Effects GLMM in Table 10.12

```
options ls = 80 ps = 60; data one; infile 'isjall.dat';
  input country gender bigvtlim jp average unemrate westeur
  postcom centeur easteur medvoc seceduc thdeduc;

proc nlmixed;
  parms gamma00 = 2.5 gamma10 = 0.5 gamma20 = 0.5 gamma30 = 0.5
gamma40 = 0.5
     su02 = 1 su12 = 0.5 su22 = 0.5 su32 = 0.5 su42 = 0.5;
  beta0 = gamma00 + u0;
  beta1 = gamma10 + u1;
  beta2 = gamma20 + u2;
  beta3 = gamma30 + u3;
  beta4 = gamma40 + u4;
  eta = beta0 + beta1*gender + beta2*medvoc + beta3*seceduc +
  beta4*thdeduc;
  expeta = exp(eta);
  p = expeta / (1 + expeta);
  model bigvtlim ~ binary(p);
  random u0 u1 u2 u3 u4
    ~normal([0,0,0,0,0],[su02,0,su12,0,0,su22,0,0,0,
    su32,0,0,0,0,su42])
    subject = country;
  run;
```

References

Alderson, Arthur S. and François Nielsen. 1999. "Income inequality, development, and dependence: a reconsideration." *American Sociological Review* 64: 606–631.

Allison, Paul D. 1984. *Event History Analysis: Regression for Longitudinal Event Data.* Beverly Hills, CA: Sage.

Anderson, Sharon, Ariane Auquier, Walter W. Hauck, David Oakes, Walter Vandaele, and Herbert I. Weisberg. 1980. *Statistical Methods for Comparative Methods: Techniques for Bias Reduction.* New York: Wiley.

Andrews, Donald W. and Ray C. Fair. 1988. "Inference in nonlinear econometric models with structural change." *Review of Economic Studies* 55: 615–640.

Arber, Sara and Helen Cooper. 1999. "Gender differences in health in later life: the new paradox?" *Social Science & Medicine* 48: 61–76.

Arts, Wil and Loek Halman (Eds.). 1999. *New Directions in Quantitative Comparative Sociology.* Boston: Brill.

Barndorff-Nielsen, O. E. 1978. *Information and Exponential Families in Statistical Theory.* Winchester: Wiley.

Barndoff-Nielsen, O. E. 1980. "Exponential families." Memoir 5. Department of Theoretical Statistics, Institute of Mathematics, University of Aarhus.

Bartholomew, David J. and Martin Knott. 1999. *Latent Variable Models and Factor Analysis.* Second Edition. Oxford: Oxford University Press.

Becker, Jerry P. (Ed.). 1992. *Report of U.S.–Japan Cross-National Research on Students' Problem Solving Behaviors.* Carbondale, IL: Southern Illinois University.

Blossfeld, H., A. Hamerle, and K. U. Mayer. 1988. *Event History Analysis: Statistical Theory and Applications in Economic and Social Sciences.* Hillsdale, NJ: Erlbaum.

Blossfeld, H. and G. Rohwer. 1995. *Techniques of Event History Modeling.* Mahwah, NJ: Lawrence Erlbaum Associates.

Bock, R. D. 1972. "Estimating item parameters and latent ability when responses are scored in two or more nominal categories." *Psychometrika* 37: 29–51.

Bollen, Kenneth A. 1989. *Structural Equations with Latent Variables.* New York: Wiley.

Bornstein, Marc H., O. Maurice Haynes, and Hiroshi Azuma. 1998. "A cross-national study of self-evaluations and attributions in parenting: Argentina, Belgium,

France, Israel, Italy, Japan, and the United States." *Developmental Psychology* 34: 662–676.

Breslow, N. E. 1982. "Covariance adjustment of relative-risk estimates in matched studies." *Biometrics* 38: 661–672.

Breslow, N. E. 1996. "Generalized linear models: Checking assumptions and strengthening conclusions." *Statistica Applicata* 8: 23–41.

Browne, Beverly. 1998. "Gender stereotypes in advertising on children's television in the 1990s: a cross-national analysis." *Journal of Advertising* 27: 83–96.

Browne, M. W. 1984. "Asymptotically distribution-free methods for the analysis of covariance structures" *British Journal of Mathematical and Statistical Psychology* 37: 62–83.

Bryk, Anthony S. and Stephen W. Raudenbush. 1996. *Hierarchical Linear Models: Applications and Data Analysis Methods*. Newbury Park, CA: Sage.

Buse, A. 1982. "The likelihood ratio, Wald, and Lagrange multiplier tests: an expository note." *The American Statistician* 36: 153–157.

Cai, Jinfa. 1995. "A cognitive analysis of U.S. and Chinese students' mathematical performance on tasks involving computation, simple problem solving, and complex problem solving." *Journal for Research in Mathematics Education Monograph Series*, No. 7. Reston, VA: National Council of Teachers of Mathematics.

Choi, Namkee G. 1997. "Racial differences in retirement income: the roles of public and private income sources." *Journal of Aging & Social Policy* 9: 21–42.

Chow, Gregory C. 1960. "Tests of equality between sets of coefficients in two linear regressions." *Econometrica* 28: 591–605.

Clogg, Clifford C. 1978. "Adjustment of rates using multiplicative models." *Demography* 15: 523–539.

Clogg, Clifford C. 1981. "New developments in latent structure analysis." Pp. 215–246 in *Factor Analysis and Measurement in Sociological Research*, edited by D. J. Jackson and E. F. Borgotta. Beverly Hills, CA: Sage.

Clogg, Clifford C. and Scott R. Eliason. 1987. "Some common problems in log-linear analysis." *Sociological Methods Research* 16: 8–44.

Clogg, Clifford C. and Scott R. Eliason. 1988. "A flexible procedure for adjusting rates and proportions, including statistical methods for group comparisons." *American Sociological Review* 53: 267–283.

Clogg, Clifford C. and Leo A. Goodman. 1984. "Latent structure analysis of a set of multidimensional contingency tables." *Journal of American Statistical Association* 79: 762–771.

Clogg, Clifford C. and Leo A. Goodman. 1985. "Simultaneous latent structure analysis in several groups." Pp. 81–110 in *Sociological Methodology* 1985, edited by Nancy B. Tuma. San Francisco: Jossey-Bass.

Clogg, Clifford C. and James W. Shockey. 1988. "Multivariate analysis of discrete data." In *Handbook of Multivariate Experimental Psychology*, edited by J. R. Nesselroade and R. B. Cattell. New York: Plenum.

Clogg, Clifford C., James W. Shockey, and Scott R. Eliason. 1990. "A general statistical framework for adjustment of rates." *Sociological Methods & Research* 19: 156–195.

Courgeau, D. and E. Lelievre. 1992. *Event History Analysis in Demography*. Oxford: Clarendon.

Cox, D. R. 1972. "Regression models and life tables." *Journal of the Royal Statistical Society B* 34: 187–203.

Cox, D. R. 1975. "Partial Likelihood." *Biometrika* 62: 269–276.

Cox, D. R. and D. Oakes. 1984. *Analysis of Survival Data*. London: Chapman and Hall.

Cressie, N. and P. W. Holland. "Characterizing the manifest probabilities of latent trait models." *Psychometrika* 48: 129–141.

Croon. M. 1990. "Latent class analysis with ordered classes." *British Journal of the Mathematical Statistical Society* 43: 171–192.

Das Gupta, Biman Kumar. 1994. "The comparative method in anthropology." *Current Anthropology* 35: 558–559.

Das Gupta, Prithwis. 1991. "Decomposition of the difference between two rates and its consistency when more than two populations are involved." *Mathematical Population Studies* 3: 125.

Demo, D. H. and K. D. Parker. 1987. "Academic achievement and self-esteem among black and white college students." *Journal of Social Psychology* 127: 345–355.

Descartes, René. 1963. *Œuvres Philosophiques*, Vol. 1, edited by Ferdinand Alqui, Paris: Garnier.

Dey, Dipak, Sunjit K. Ghosh, and Bani K. Mallick. 2000. *Generalized Linear Models: A Bayesian Perspective*. New York: Marcel Dekker.

Dobson, Annette J. 1990. *An Introduction to Generalized Linear Models*. London: Chapman and Hall

Echevarria, Cristina and Antonio Merlo. 1999. "Gender differences in education in a dynamic household bargaining model." *International Economic Review* 40: 265–286.

Efron, Bradley, and Robert J. Tibshirani. 1993. *An Introduction to the Bootstrap*. New York: Chapman and Hall.

Ehrhardt-Martinez, Karen. 1998. "Social determinants of deforestation in developing countries: a cross-national study." *Social Forces* 77: 567–586.

Ellina, Maro and Will H. Moore. 1990. "Discrimination and political violence: a cross-national study with two time periods." *Western Political Quarterly* 43: 267–278.

Fisher, R. A. 1932. *Statistical Methods for Research Workers*. Fourth Edition. Edinburgh: Oliver and Boyd.

Fisher, R. A. 1935. *The Design of Experiments*. First Edition. Edinburgh and London: Oliver and Boyd.

Fleiss, Joseph L. 1981. *Statistical Methods for Rates and Proportions*. Second Edition. New York: Wiley.

Fleiss, Joseph L. 1986. *The Design and Analysis of Clinical Experiments*. New York: Wiley.

Gershuny, Jonathan. 2000. *Changing Times: Work and Leisure in Postindustrial Society*. Oxford: Oxford University Press.

Goethe, Johann Wolfgang von. 1982. *Werke: Hamburger Ausgabe*, Vol. 13. Munich: DTV.

Goldstein, H. 1995. *Multilevel Statistical Models*. London: Edward Arnold.

Goldstein, H., J. Rasbash, W. Browne, G. Woodhouse, and M. Poulain. 2000. "Multilevel models in the study of dynamic household structures." *European Journal of Population* 16: 373–387.

Good, Phillip. 2000. *Permutation Tests: A Practical Guide to Resampling Methods for Testing Hypotheses*. Second Edition. New York: Springer.

Goodman, Leo A. 1973. "The analysis of multidimensional contingency tables when some variables are posterior to others: a modified path analysis approach." *Biometrika* 60: 179–192.

Greene, William H. 1990. *Econometric Analysis*. New York: Macmillan.

Haberman, Shelby J. 1979. *Analysis of Qualitative Data, Vol. 2: New Developments*. New York: Academic.

Hagenaars, Jacques A. 1988. "Latent structure models with direct effects between indicators: local dependence models." *Sociological Methods Research* 16: 379–405.

Hagenaars, Jacques A. 1993. *Loglinear Models with Latent Variables*. Newbury Park, CA: Sage.

Handcock, Mark S. and Martina Morris. 1998. "Relative distribution methods." Pp. 53–97 in *Sociological Methodology* 1998, edited by Adrian E. Raftery. Boston: Blackwell.

Handcock, Mark S. and Martina Morris. 1999. *Relative Distribution Methods in the Social Sciences*. New York: Springer-Verlag.

Hastie, T. J. and R. J. Tibshirani. 1990. *Generalized Additive Models*. London: Chapman and Hall.

Heinen, Ton. 1993. *Discrete Latent Variable Models*. Tilburg: Tilburg University Press.

Henriques, Gregg R. and Lawrence G Calhoun. 1999. "Gender and ethnic differences in the relationship between body esteem and self-esteem." *The Journal of Psychology* 133: 357–368.

Higgins, Joan. 1981. *States of Welfare: Comparative Analysis in Social Policy*. Oxford: Blackwell and Martin Robertson.

Hill, P. W. and H. Goldstein. 1998. "Multilevel modelling of educational data with cross classification and missing identification of units." *Journal of Educational Behavioural Statistics* 23: 117–128.

Honda, Yuzo and Kazuhiro Ohtani. 1986. "Modified Wald tests in tests of equality between sets of coefficients in two linear regressions under heteroscedasticity." *The Manchester School of Economic and Social Studies* 54: 208–218.

Hox, J. J. 1994. *Applied Multilevel Analysis*. Amsterdam: TT-Publikaties.

Hupkens, Christianne L. H., Ronald A. Knibbe, and Anneke H. van Otterloo. 1998. "Class differences in the food rules mothers impose upon their children: a cross-national study (Netherlands, Belgium, and Germany)." *Social Science & Medicine* 47: 1331–1339.

Ishida, Hiroshi, Walter Muller, and John M Ridge. 1995. "Class origin, class destination, and education: a cross-national study of ten industrial nations." *American Journal of Sociology* 101: 145–193.

Johnson, Monica Kirkpatrick and Margaret Mooney Marini. 1998. "Bridging the racial divide in the United States: the effects of gender." *Social Psychology Quarterly* 61: 247–258.

Kalbfleisch, J. D. and R. L. Prentice. 1980. *The Statistical Analysis of Failure Time Data*. New York: Wiley.

Kass, Robert E. and Adrian E. Raftery. 1995. "Bayes factor." *Journal of the American Statistical Association* 90: 773–795.

Kenworthy, Lane. 1999. "Do social-welfare policies reduce poverty? A cross-national assessment." *Social Forces* 77: 1119–1139.

King, Gary. 1989. *Unifying Political Methodology: The Likelihood Theory of Statistical Inference*. Cambridge: Cambridge University Press.

Kitagawa, E. M. 1955. "Components of a difference between two rates." *Journal of the American Statistical Association* 50: 1168–1194.

Kluegel, James R., David S. Mason, and Bernd Wegener (Eds.). 1995. *Social Justice and Political Change: Public Opinion in Capitalist and Post-Communist States*. Hawthorne, NY: Aldine de Gruyter.

Kohn, Melvin L. (Ed.). 1989. *Cross-National Research in Sociology*. American Sociological Association Presidential Series. Newbury Park, CA: Sage.

Krall, J. M., V. A. Uthoff, V. A. and J. B. Harley. 1975. "A step-up procedure for selecting variables associated with survival." *Biometrics* 31: 49–57.

Kramer, C. Y. 1956. "Extension of multiple range test to group means with unequal number of replications." *Biometrics* 12: 307–310.

Kreft, I. G. and J. de Leeuw. 1998. *Introducing Multilevel Modeling*. Thousand Oaks, CA: Sage.

Lacour, Claudia Brodsky. 1995. "Grounds of comparison." *World Literature Today* 69: 271–274.

Lacy, Michael G. 1997. "Efficiently studying rare events: case–control methods for sociologists." *Sociological Perspectives* 40: 129–154.

Lancaster, T. 1990. *The Economic Analysis of Transition Data*. Cambridge: Cambridge University Press.

Lawless, J. F. 1982. *Statistical Models and Methods for Lifetime Data*. New York: Wiley.

Lazarsfeld, Paul F. and Neil W. Henry. 1968. *Latent Structure Analysis*. Boston: Houghton Mifflin.

Lesaffre, E. and B. Spiessens. 2001. "On the effect of the number of quadrature points in a logistic random-effects model: An Example." *Journal of the Royal Statistical Society Series C—Applied Statistics* 50 (Part 3): 325–335.

Levene, H. 1960. "Robust tests for equality of variances." Pp. 278–292 in *Contributions to Probability and Statistics: Essays in Honor of Harold Hotelling*, edited by I. Olkin et al. Stanford, CA: Stanford University Press.

Liao, F. 1989a. "A flexible approach for the decomposition of rate differences." *Demography* 26: 717–726.

Liao, F. 1989b. "Fertility differentials of religious and ethnic groups in the United States: a fuzzy group membership approach." Ph.D. Dissertation. University of North Carolina at Chapel Hill.

Liao, F. 1989c. "Estimating fuzzy set membership coefficients with log-multiplicative association models: the case of contraceptiveness." *Mathematical Population Studies* 1: 357–376.

Liao, T. F. 2002. "Bayesian model comparison in generalized linear models across multiple groups." *Computational Statistics and Data Analysis* 39.

Lloyd, Kim M. and Scott J. South. 1996. "Contextual influences on young men's transition to first marriage." *Social Forces* 74: 1097–1119.

Longford, N. T. 1993. *Random Coefficient Models*. New York: Oxford University Press.

Lorenz, Max O. 1905. "Methods of measuring the concentration of wealth." *Journal of the American Statistical Association* 9: 209–219.

Mabbett, Deborah and Helen Bolderson. 1999. "Theories and Methods in Comparative Social Policy." Pp. 34–56 in *Comparative Social Policy: Concepts, Theories and Methods*, edited by Jochen Clasen. Oxford: Blackwell.

Manton, K. G., K. Liu, and E. S. Cornelius. 1985. "An analysis of the heterogeneity of U.S. nursing home patients." *Journal of Gerontology* 40: 34–46.

Manton, K. G., E. Stallard, M. A. Woodbury, H. D. Tolley, and A. I. Yashin. 1987. "Grade-of-membership techniques for studying complex event history processes with unobserved covariates." Pp. 309–346. *Sociological Methodology* 1987, edited by C. C. Clogg. Washington, DC: American Sociological Association.

Martin, William G. and Mark Beittel. 1998. "Toward a global sociology? Evaluating current conceptions, methods, and practices." *Sociological Quarterly* 39: 139–161.

McCullagh, P. and J. A. Nelder. 1989. *Generalized Linear Models*. Second Edition. London: Chapman and Hall.

McCulloch, C. E. 1997. "Maximum likelihood algorithms for generalized linear mixed models." *Journal of the American Statistical Association* 92: 162–170.

McDowell, John M., Larry D. Singell Jr., and James P. Ziliak. 1999. "Cracks in the glass ceiling: gender and promotion in the economics profession." *American Economic Review* 89: 392–396.

McMichael, Philip. 1990. "Incorporating comparison within a world-historical perspective: an alternative comparative method." *American Sociological Review* 55: 385–397.

McNemar, Q. 1947. "Note on the sampling error of the difference between correlated proportions or percentages." *Psychometrika* 12: 153–157.

Mennen, Ferol E. 1994. "Sexual abuse in Latina girls: their functioning and a comparison with white and African American girls." *Hispanic Journal of Behavioral Sciences* 16: 475–486.

Miller, John William. 1981. *The Philosophy of History with Reflections and Aphorisms*. New York: Norton.

Mitchell, W. J. T. 1996. "Why comparisons are odious." *World Literature Today* 70: 321–324.

Moustaki, I and M. Knott. 2000. "Generalized latent trait models." *Pychometrika* 65: 391–411.

Mueller, Ralph O. 1996. *Basic Principles of Structural Equation Modeling: An Introduction to LISREL and EQS*. New York: Springer.

Nelder, J. A. 1998. "A large class of models derived from generalized linear models." *Statistics in Medicine* 17: 2747–2753.

Nelder, J. A. and R. W. M. Wedderburn. 1972. "Generalized linear models." *Journal of the Royal Statistical Society A* 135: 370–384.

Nowak, Stefan. 1989. "Comparative studies and social theory." Pp. 34–56 in *Cross-National Research in Sociology*, edited by Melvin L. Kohn. American Sociological Association Presidential Series. Newbury Park, CA: Sage.

Otterbein, Keith F. 1994. "The comparative method in anthropology." *Current Anthropology* 35: 559–560.

Perrucci, Carolyn C., Robert Perrucci, and Dena B. Targ. 1997. "Gender differences in the economic, psychological and social effects of plant closings in an expanding economy." *Social Science Journal* 34: 217–233.

Peters, B. Guy. 1998. *Comparative Politics: Theory and Methods*. New York: New York University Press.

Raftery, Adrian. 1995. "Bayesian model selection in social research." Pp. 111–163 in *Sociological Methodology* 1995, edited by Peter V. Marsden. Cambridge, MA: Blackwell.

Raftery, Adrian. 1996. "Approximate Bayes factors and accounting for model uncertainty in generalised linear models." *Biometrika* 83: 251–266.

Rao, J. N. K. and A. J. Scott. 1992. "A simple method for the analysis of clustered binary data." *Biometrics* 48: 577–586.

Rao, J. N. K. and A. J. Scott. 1999. "A simple method for analysing overdispersion in clustered Poisson data." *Statistics in Medicine* 18: 1373–1386.

Ridley, Mark. 1994. "The comparative method in anthropology." *Current Anthropology* 35: 560–561.

Rigby R. A. and M. D. Stasinopoulos 1995. "Mean and dispersion additive models: applications and diagnostics." Pp. 249–256 in *Statistical Modelling: Proceedings of the 10th International Workshop on Statistical Modelling*, edited by G. U. H. Seeber, B. J. Francis, R. Hatzinger, and G. Steckel-Berger. New York: Springer-Verlag.

Rigby R. A. and M. D. Stasinopoulos 1996. "A semi-parametric additive model for variance heterogeneity." *Statistics and Computing* 6: 57–65.

Sahai, Hardeo and Mohammed I. Ageel. 2000. *The Analysis of Variance: Fixed, Random, and Mixed Models*. Boston: Birkhauser.

Sapiro, Virginia and Pamela Johnston Conover. 1997. "The variable gender basis of electoral politics: gender and context in the 1992 US election." *British Journal of Political Science* 27: 497–523.

Schall, Robert. 1991. "Estimation in generalized linear models with random effects." *Biometrika* 78: 719–727.

Scheffé, H. 1953. "A method for judging all contrasts in the analysis of variance." *Biometrika* 40: 87–104.

Scheffé, H. 1959. *The Analysis of Variance*. New York: Wiley.

Schweizer, Thomas. 1994. "The comparative method in anthropology." *Current Anthropology* 35: 561.

Silver, Edward A., S. S. Leung, and Jinfa Cai. 1995. "Generating multiple solutions for a problem: a comparison of the responses of U.S. and Japanese students." *Educational Studies in Mathematics* 28: 35–54.

Smith, Herbert L. 1997. "Matching with multiple controls to estimate treatment effects in observational studies." Pp. 325–353 in *Sociological Methodology 1997*, edited by Adrian E. Raftery. Boston: Blackwell.

Smith, Philip J. and Daniel F. Heitjan. 1993. "Testing and adjusting for departures from nominal dispersion in generalized linear models." *Applied Statistics* 42: 31–41.

Smyth, G. K. 1989. "Generalized linear models with varying dispersion." *Journal of Royal Statistical Society B*, 51: 47–60.

Smyth, G. K. and A. P. Verbyla, 1996. "A conditional approach to residual maximum likelihood estimation in generalized linear models. *Journal of Royal Statistical Society B*, 58: 565–572.

Smyth, G. K. and A. P. Verbyla, 1999. "Adjusted likelihood methods for modelling dispersion in generalized linear models." *Environmetrics* 10: 695–709.

Snijders, T. and R. Bosker. 1999. *Multilevel Analysis: An Introduction to Basic and Advanced Multilevel Modeling*. Thousand Oaks, CA: Sage.

Sörbom, D. 1981. "Structural equation models with structured means." In *Systems under Indirect Observation: Causality, Structure and Prediction*, Vol. 1, edited by K. G. Jreskog and H. Wold. Amsterdam: North-Holland Publishing Co.

Sullivan, J. L. and J. E. Transue. 1999. "The psychological underpinnings of democracy: a selective review of research on political tolerance, interpersonal trust, and social capital." *Annual Review of Psychology* 50: 625–650.

Telles, Edward E. and Nelson Lim. 1998. "Does it matter who answers the race question? Racial classification and income inequality in Brazil." *Demography* 35: 465–474.

Tukey, John W. 1953. "The problem of multiple comparisons" (Mimeograph, 396 pages). Princeton, NJ: Department of Mathematics, Princeton University.

Tuma, Nancy B. and Michael T. Hannan. 1984. *Social Dynamics: Models and Methods*. New York: Academic.

Vermunt, Jeroen K. 1997. *Log-Linear Models for Event Histories*. Thousand Oaks, CA: Sage.

Voland, Eckart. 1994. "The comparative method in anthropology." *Current Anthropology* 35: 561–562.

Wald, Abraham. 1943. "Tests of statistical hypothesis concerning several parameters when the number of observations is large." *Transactions of the American Mathematical Society* 54: 426–482.

Wesolowski, Wlodzimierz. 1989. "Legitimate domination in comparative-historical perspective: the case of legal domination." Pp. 34–56 in *Cross-National Research in Sociology*, edited by Melvin L. Kohn. American Sociological Association Presidential Series. Newbury Park, CA: Sage.

Westholm, Anders and Richard G. Niemi. 1992. "Political institutions and political socialization: a cross-national study." *Comparative Politics* 25: 25–41.

Wonnacott, Ronald J. and Thomas H. Wonnacott. 1979. *Econometrics*. Second Edition. New York: Wiley.

Wright, Erik Olin, Janeen Baxter, and Gunn Elizabeth Birkelund. 1995. "The gender gap in workplace authority: a cross-national study." *American Sociological Review* 60: 407–435.

Yamaguchi, Kazuo. 1991. *Event History Analysis*. Newbury Park, CA: Sage.

Index

Analysis of covariance (ANCOVA), 6, 21, 30, 36, 178. *See also* Multilevel analysis
Analysis of variance (ANOVA), 5, 6, 21, 25–26, 29, 36, 178. *See also* Multilevel analysis
Association, 20, 56, 146, 156–157, 161–162
Assumption
   association, 153
   distributional, 37, 122
   for the Chow test, 31
   for linear model, 23
   local independence, 145–146
   normal distribution (normality), 37, 39, 130
   of dispersion homogeneity, 88, 90, 95
   of homoscedasticity, 24
   of linear multilevel model, 186
   sampling, 151
Asymptotic equivalent, 6
Asymptotic test, 16
Asymptotically distribution free (ADF) method, 117
Asymptotically normally distributed, 18

Bartlett's test, *see* Test statistic
Baseline hazard function, 86
Bayes factor, 98–100, 125
Bayes theorem, 97–98
Bayesian
   approach, 5, 101
   generalized linear model, 96–110
   inference, 96–104
   information criterion (BIC), 99, 101, 103, 105, 122–123
   statistics, 7, 96
Bias, 5
Bias reduction, 5
Binary outcome, 65
Binomial distribution, *see* Distribution
Bonferroni inequality, 29

Bonferroni method, *see* Multiple comparison method
Bootstrap distribution, *see* Distribution
Bootstrapping method, 42–44, 91

Canonical link, 66, 88
Canonical parameter, 66, 87
Case-control design (study), 81–87
   1:1, 82–84
   $1:m$, 82–85
   $n:m$, 82–83, 85–87, 110
Categorical latent variable, 7–8, 142, 144
Censoring, 75
Chi-square distribution, *see* Distribution
Chi-square statistic, *see* Test statistic
χ^2-test, *see* Test statistic
Chow test, *see* Test statistic
Clinical trials, 82, 174
Clustered data, 88, 174
Cochran's test, *see* Test statistic
Comparison
   and multilevel analysis, 191–193
   Bayesian model, 98–107, 125
   cross-cultural, 4, 172
   cross-national, 3–4
   in the social sciences, 3–5
   of baseline hazard function, 78–80
   of cases and controls, 85–87
   of categorical latent variables, 151–154
      association, 152–153
      conditional probability, 153, 157–158
      latent class analysis, 154–158
      latent probability, 153–154, 157–158
      path model, 158–161
      sampling distribution, 152, 156–157
   of data structure, 19
   of dispersion, 94–95
   of distribution, 19, 172

Comparison (*Continued*)
  of generalized linear models, 68–80
  of intercept, 20, 34
  of means, 4, 6, 9, 20, 24–25, 27–30, 34, 41–42
  of model structure, 19, 172
  of parameter (coefficient), 19, 20, 34, 71–74, 79, 94, 100–101, 104–107, 172
  of proportions, 4, 6, 24
  of rates, 4, 6, 24, 51–61
  of residual mean, 20, 35
  of response variable, 20
  of structural equation model, 121–127
    covariance matrix, 123–124, 127–129
    model form, 124–125
    multivariate distribution, 122–123, 130–131
    model parameter, 125–127, 130–132
      covariance structure, 130–132, 132–134
      general model, 126
      mean structure, 126–127, 132–134
      measurement model, 125–126
      structural model, 125
  philosophy for, 1–3
Comparative research, 3
Complementary log-log link, *see* Link function
Component effect, 59
Composition factor, 51
Composition structure, 54
Computer software
  Amos, 111
  BMDP, 196
  BUGS, 197
  CDAS, 57, 161
  `decompos`, 61
  DNEWTON, 161
  EQS, 7, 111–112, 117, 120, 128, 134–140
  `glib`, 103
  Glim, 92
  HLM, 188, 195
  LAT, 161
  Latent GOLD, 162
  `lca.S`, 163
  `lca.sas`, 163
  LCAG, 162
  LEM, 156, 160, 162–171
  LISCOMP, 163
  LISREL, 7, 111–112, 117, 120, 128, 130, 136, 138–141, 197
  MIXREG, MIXOR, MIXNO, MIXPREG, MIXGSUR, 195
  MLLSA, 161
  MLn, MLwiN, 196
  Mplus, 111, 163, 197
  NEWTON, 161
  `multigroup.glib`, 103
  PANMARK, 161–162
  S(-plus), 40, 47, 86, 89, 92, 96, 103, 161, 196
  SAS, 43, 47, 85–86, 91, 161, 163, 173, 188, 189, 192, 195–198
  SPSS, 38, 39, 43, 84, 86, 162, 192
  Stata, 42, 43, 85, 91, 163, 196
  VARCL, 195
Concomitant variable, 30
Conditional likelihood, 82
Conditional logistic regression, 82–87
Confidence interval, 28–29
Confidence level, 14–15, 43
Confounding factor, 5, 51, 61, 82, 85
Confounding variable, 5, 36
Cohort study, 82
Constraint, 22, 28, 79, 121, 124, 126, 132–133, 145, 160, 162, 172, 178–179, 190
Contextual analysis, 173
Contingency table, 19, 53, 83, 146, 151–152
Control group, 7
Convergence, 67, 161
Correlation matrix, 19, 123
Covariance structure, 7
Covariance structure model, 112–118, 120–121
Cox's model, 77–80, 86. *See also* Hazard rate model; Proportional-hazards model
Cressie-Reed statistic, 157–158, 160

Decomposition, 51, 58–61
  arithmetic, 58–59
  model-based, 59–61
Degree(s) of freedom, 11–12, 15, 17, 26, 28, 31, 69, 71, 73, 105, 185
Discrete-time model, 77–78
Dispersion
  heterogeneity, 87–96
  homogeneity, 70, 90, 94
  parameter, 63, 81, 87–88, 94
  submodel, 92, 95–96
Distribution
  binomial, 64–65, 88, 148, 195
  bootstrap, 42
  chi-square, 10–11, 15, 18
  exact permutation, 41
  exponential, 64
  extreme value, 64, 66
  gamma, 64, 195
  Gaussian, 63
  inverse Gaussian, 64
  joint, 144, 148

latent, 151
logistic, 65, 148
marginal, 149
multinomial, 66, 148, 151
negative binomial, 64, 196
normal, 23, 31, 63, 65, 88, 122, 186, 195
Pareto, 64
permutation, 42
Poisson, 64, 65, 88, 151, 195–196
product-multinomial, 151
relative, 44–45, 47, 50
sampling, 152
tolerance, 65
Double generalized linear model (DGLM), 91–96
Duration analysis, 74. *See also* Hazard rate model

Ecological fallacy, 174
EM (algorithm), 149, 161, 163
Endogenous variable, 112–114, 121
Equality hypothesis, 14
Event history analysis, 74, 162. *See also* Hazard rate model
Exact permutation, *see* Distribution
Exogenous variable, 112–114, 121, 125, 126
Experiment, 7, 25, 30, 36, 81, 121, 174
Exponential distribution, *see* Distribution
Exponential family, 62–63, 148
Extreme value distribution, *see* Distribution

F density, 14
F-distribution, 13, 28, 71
F-ratio, 13
F-test, *see* Test statistic
Failure-time analysis, 74
Fisher's information matrix, *see* Information matrix
Fixed component, 177, 190, 196
Fixed effects, 21, 172–173, 175–176, 183, 190–193
Frequentist approach, 5
Fuzzy set theory, 194

Gamma distribution, *see* Distribution
Gaussian distribution, *see* Distribution
Gaussian quadrature, 173, 188, 196
Generalized addition rule, 97
Generalized additive model (GAM), 92
Generalized estimating equations (GEE), 196
Generalized least squares (GLS), 116–117
Generalized linear latent and mixed model (GLLAMM), 186, 196

Generalized linear mixed model, 5, 173, 186–190, 192–194
Generalized linear model (GLM), 5–8, 19, 22, 55, 62–80, 81–110, 172, 174, 183–184, 186
Generalized linear multilevel model (GLMM), 186–190, 192–194
Generalized multiplication rule, 96
Gini coefficient, 45–47
Grouped data, 90

Hartley's test, *see* Test statistic
Hazard rate, 76
Hazard rate model, 6, 74–80
  parametric, 76
    exponential, 76
    gamma, 76
    Gompertz, 76
    inverse-Gaussian, 76
    log-logistic, 76
    log-normal, 76
    piecewise exponential, 76
    Weibull, 76
  nonparametric, 77
  semiparametric
    Cox's proportional hazard model, 77–80, 86
Heteroscedasticity, 7, 88
Hierarchical (linear) model, 173–175, 197. *See also* Multilevel analysis
Homogeneity, 30
Homoscedasticity, 70–71, 88

Identification, *see* Structural equation model
Identity link, *see* Link function
Independently and identically distributed (i.i.d.), 62, 174
Information matrix, 67–68, 70, 91
Interaction, 31, 56, 61, 72–73, 76, 94, 101, 103, 145–146, 148, 150, 156–157, 172, 179, 184, 191
Intraclass correlation, 177, 181–182
Inverse Gaussian distribution, *see* Distribution
Item response function, 147
Item response theory, 148
Iterative reweighted least squares (IRLS), 67

Jackknife, 91
Joint distribution, *see* Distribution

Kolmogorov-Smirnov test, *see* Test statistic

Lagrange multiplier statistic, 6
Lagrange multiplier test (LMT). *See also* Test statistic

Lagrange multiplier test (*Continued*)
  in conditional logistic regression, 87
  in logit model, 73
  in structural equation model, 128–129, 132–133
Laplace method, 99
Latent class model (analysis) (LCM), 8, 143–149, 162–163
Latent distribution, *see* Distribution
Latent profile model (LPM), 143, 148–149
Latent structure model, 142, 161–163
Latent trait model (LTM), 143, 147–149, 162–163
Latent variable mean, 120
Latent variable model, 9, 111–134, 142–161
Least significance difference (LSD) test, *see* Multiple comparison method
Levene's test, *see* Test statistic
Likelihood equation, 67
Likelihood function, 67, 82
Likelihood ratio statistic, 6
Likelihood ratio test (LRT). *See also* Test statistic
  in conditional logistic regression, 87
  in hazard rate model, 78–80
  in latent class model, 157–158, 160
  in logit model, 73
  in multilevel analysis, 184
  in structural equation model, 132–133
Linear growth rate, 22
Linear logit model, 72
Linear model, 5, 21–36, 111, 196
Linear predictor, 63
Linear regression, 6, 40, 62, 71, 88, 111, 122, 178. *See also* Multiple linear regression
Link function, 19, 62–66
  complementary log–log, 66
  identity, 65
  logarithm, 65
  logit, 65, 77, 104, 163, 196
  multinomial logit, 65
  nonlinear, 71
  nonparametric, 92
  parametric, 92
  probit, 65, 163, 196
Local dependence model, 146–147, 162
Local independence, 146. *See also* Assumption
Location parameter, 24, 44
Logarithm link, *see* Link function
Logistic distribution, *see* Distribution
Logistic-ogive model, 147
Logistic regression, 71–74, 83–85
Logit link, *see* Link function
Logit model (analysis), 5–6, 62, 71–74, 77

Log-likelihood function, 67, 69, 79
Loglinear model (analysis) 5, 8, 55–56, 62, 76, 145–146, 149–150, 152–153, 161–162
Lognormal distribution, 19
Lorenz curve, 45–47

Mann-Whitney test, *see* Test statistic
McNemar test, *see* Test statistic
Markov chain model, 162
Matched studies, 7, 81–87. *See also* Case-control design
Matching, 5
Maximum likelihood (ML), 14, 116–117, 185
Maximum likelihood estimation (MLE), 66–68, 196
Mean and dispersion additive model (MADAM), 92
Mean structure, 7
Mean submodel, 92, 95–96
Measurement error, 113–114, 126
Measurement model, 20, 112–114, 172
MIMIC model, 149, 153, 163
Mixed-effects model, 173–174, 195–196
Model space, 100–103, 110
Monte Carlo simulation, 39
Multiequation system, 7
Multilevel analysis, 8, 172–198
  group as external, 190
  multiple membership, 193–194
  random coefficient logit with cross-level effects, 187–189
  random coefficient model, 179–180
    with cross-level effects, 179–180, 183–185
    without cross-level effects, 179, 183
  random coefficient probit with cross-level effects, 190
  random intercept model, 177–178
    one-way ANCOVA, 177–178, 181–183
    one-way ANOVA, 177, 181
Multinomial distribution, *see* Distribution
Multinomial logit link, *see* Link function
Multinomial logit model, 6, 195
Multiple classification analysis (MCA), 33
Multiple comparison method, 27–30
  Bonferroni's method, 29–30
  least significance difference (LSD) test, 27
  Scheffé's method, 28–29
  Tukey's method, 27–28
Multiple groups, 5, 27–28, 70, 100, 151
Multiple group analysis, 19
Multiple linear regression, 30–36
Multivariate elliptical distribution, 117, 122, 130
Multivariate normal distribution, 122, 130

INDEX   **211**

Negative binomial distribution, *see* Distribution
Nonlinear link, *see* Link function
Nonlinear regression, 43
Nonparametric comparison, 37–50
Nonparametric link, *see* Link function
Nonparametric method, 5–6, 23
Nonparametric test, 9–10, 24, 37–39
Normal distribution, *see* Distribution
Normal ogive, 147
Nuisance parameter, 82

Odds ratio, 22
Offset, 84
Ordinary least squares (OLS), 23
Overdispersion, 88–89, 94

Parametric distribution, 9. *See also* Distribution
Parametric link, *see* Link function
Parametric method, 5, 6, 23, 37
Parametric model, 66, 78
Pareto distribution, *see* Distribution
Partial likelihood, 77, 86
Path model, 8, 114, 162, 197
   with categorical latent variables, 149–151
Pearson chi-square, 11
Pearson deviance statistic, *see* Test statistic
Permutation
   distribution, *see* Distribution
   method, 37, 39–42
   test, 9, 10, 39–42
Poisson distribution, *see* Distribution
Poisson regression, 62, 195
Pooling of time series and cross-sectional data, 173
Posterior odds, 98, 102–103, 106–108, 125
Posterior probability, 96–98, 100, 102–103, 106–108
Prior distribution, 105
Prior odds, 98, 100, 107
Prior probability, 96–97, 102, 110
Prior weight, 87–89, 110
Probit link, *see* Link function
Probit model, 6, 62
Product-multinomial distribution, *see* Distribution
Proportional-hazards model, 77–80, 86. *See also* Hazard rate model
Prospective study, 82
Purging method, 56–61, 161

Quadratic function (form or measure), 18, 99
Quasiexperiment, 7, 26, 36, 121

Quasilikelihood, 90–91, 94–95
Quasi-likelihood-ratio, 7

Random component, 62–63, 177, 190, 196
Random distribution, 6, 8
Random effects, 21, 172–173, 175–177, 190–193
Randomization test, 39, 41
Rasch model, 147
Reference group, *see* Standard group
Regression decomposition, 6, 33–36
Relative distribution, *see* Distribution
Relative distribution method, 6, 19, 37, 44–50
Relative risk, 54–55
Resampling method, 6, 39–44, 91
Restriction, 14, 146, 148, 150
Retrospective study, 82
Risk period, 75
Risk set, 75, 86

Sampling distribution, *see* Distribution
Sample size, 29, 33
Sampling zero, 19
Sandwich variance estimator, 91
Scheffé's method, *see* Multiple comparison method
Schwarz's criterion, 99
Score statistic, *see* Test statistic
Single-step Newton method, 105
Software, *see* Computer software
Standard (reference) group, 35, 44, 50, 52–53, 60–61
Standard mortality ratio, 54–55
Standard normal distribution, 12
Standardization, 5, 51–57
   direct, 52–54, 57
   indirect, 52, 54–55, 57
   model-based, 55–57
Standardized rate
   directly, 53
   indirectly, 55
Structural equation model, 111–141
   direct effect, 118
   estimation, 112, 116–117
   identification, 112, 115–116
   indirect effect, 118
   interpretation, 112, 118
   modification, 112, 117–118
   specification, 112–115
   total effect, 118
Structural means model, 119–121
Structural model, 112–114, 172
Studentized range distribution, 27
Sufficient statistic, 66

Survival analysis, 74, 195
Survival function, 75
Systematic component, 63

t-distribution, 11
t-test, *see* Test statistic
Test statistic, 6
    Bartlett's, 23
    Chow's, 31–32, 70–71
    Cochran's, 23
    χ^2, 6, 9–11, 89, 105, 122–123
    F, 6, 9, 13–14, 26–27, 30–32, 37, 89–90, 94
    Hartley's, 23
    Komogorov-Smirnov's, 37–38
    Lagrange multiplier, 6, 11, 17–18, 69–70, 73, 128, 132–133
    Levene's, 23
    likelihood ratio, 6, 11, 14–17, 68–70, 73, 89–91, 103, 105, 128, 132–133, 152–153, 160, 184–185
    Mann-Whitney's, 37–39
    McNemar's, 37, 82–83
    Pearson deviance, 89–90
    score, 89
    t, 6, 9, 12–13, 16, 24, 31, 37, 51
    Wald, 6, 11, 15–16, 69–70, 73, 91, 156

Time varying covariate, 75
Tolerance distribution, *see* Distribution
Treatment group, 7, 82–83
Type I error, 29, 43
Tukey's method, *see* Multiple comparison method

Uncertainty, 100, 110
Underdispersion, 88–89

Variance component, 26

Wald statistic. *See also* Test statistic
    in conditional logistic regression, 87
    in hazard rate model, 79
    in latent class model, 156
    in logit model, 73
Weighting, 33, 60, 90, 107
Wilcoxon test, *see* Test statistic

Z-ratio (score), 10, 12
Zero (cell count)
    sampling, 152–153, 156
    structural, 152

WILEY SERIES IN PROBABILITY AND STATISTICS
ESTABLISHED BY WALTER A. SHEWHART AND SAMUEL S. WILKS

Editors
David J. Balding, Peter Bloomfield, Noel A. C. Cressie, Nicholas I. Fisher, Iain M. Johnstone, J. B. Kadane, Louise M. Ryan, David W. Scott, Adrian F. M. Smith, Jozef L. Teugels
Editors Emeriti: *Vic Barnett, J. Stuart Hunter, David G. Kendall*

The *Wiley Series in Probability and Statistics* is well established and authoritative. It covers many topics of current research interest in both pure and applied statistics and probability theory. Written by leading statisticians and institutions, the titles span both state-of-the-art developments in the field and classical methods.

Reflecting the wide range of current research in statistics, the series encompasses applied, methodological and theoretical statistics, ranging from applications and new techniques made possible by advances in computerized practice to rigorous treatment of theoretical approaches.

This series provides essential and invaluable reading for all statisticians, whether in academia, industry, government, or research.

ABRAHAM and LEDOLTER · Statistical Methods for Forecasting
AGRESTI · Analysis of Ordinal Categorical Data
AGRESTI · An Introduction to Categorical Data Analysis
AGRESTI · Categorical Data Analysis
ANDĚL · Mathematics of Chance
ANDERSON · An Introduction to Multivariate Statistical Analysis, *Second Edition*
*ANDERSON · The Statistical Analysis of Time Series
ANDERSON, AUQUIER, HAUCK, OAKES, VANDAELE, and WEISBERG ·
 Statistical Methods for Comparative Studies
ANDERSON and LOYNES · The Teaching of Practical Statistics
ARMITAGE and DAVID (editors) · Advances in Biometry
ARNOLD, BALAKRISHNAN, and NAGARAJA · Records
*ARTHANARI and DODGE · Mathematical Programming in Statistics
*BAILEY · The Elements of Stochastic Processes with Applications to the Natural
 Sciences
BALAKRISHNAN and KOUTRAS · Runs and Scans with Applications
BARNETT · Comparative Statistical Inference, *Third Edition*
BARNETT and LEWIS · Outliers in Statistical Data, *Third Edition*
BARTOSZYNSKI and NIEWIADOMSKA-BUGAJ · Probability and Statistical Inference
BASILEVSKY · Statistical Factor Analysis and Related Methods: Theory and
 Applications
BASU and RIGDON · Statistical Methods for the Reliability of Repairable Systems
BATES and WATTS · Nonlinear Regression Analysis and Its Applications
BECHHOFER, SANTNER, and GOLDSMAN · Design and Analysis of Experiments for
 Statistical Selection, Screening, and Multiple Comparisons
BELSLEY · Conditioning Diagnostics: Collinearity and Weak Data in Regression
BELSLEY, KUH, and WELSCH · Regression Diagnostics: Identifying Influential
 Data and Sources of Collinearity
BENDAT and PIERSOL · Random Data: Analysis and Measurement Procedures,
 Third Edition

*Now available in a lower priced paperback edition in the Wiley Classics Library.

BERRY, CHALONER, and GEWEKE · Bayesian Analysis in Statistics and Econometrics: Essays in Honor of Arnold Zellner
BERNARDO and SMITH · Bayesian Theory
BHAT · Elements of Applied Stochastic Processes, *Second Edition*
BHATTACHARYA and JOHNSON · Statistical Concepts and Methods
BHATTACHARYA and WAYMIRE · Stochastic Processes with Applications
BILLINGSLEY · Convergence of Probability Measures, *Second Edition*
BILLINGSLEY · Probability and Measure, *Third Edition*
BIRKES and DODGE · Alternative Methods of Regression
BLISCHKE AND MURTHY · Reliability: Modeling, Prediction, and Optimization
BLOOMFIELD · Fourier Analysis of Time Series: An Introduction, *Second Edition*
BOLLEN · Structural Equations with Latent Variables
BOROVKOV · Ergodicity and Stability of Stochastic Processes
BOULEAU · Numerical Methods for Stochastic Processes
BOX · Bayesian Inference in Statistical Analysis
BOX · R. A. Fisher, the Life of a Scientist
BOX and DRAPER · Empirical Model-Building and Response Surfaces
*BOX and DRAPER · Evolutionary Operation: A Statistical Method for Process Improvement
BOX, HUNTER, and HUNTER · Statistics for Experimenters: An Introduction to Design, Data Analysis, and Model Building
BOX and LUCEÑO · Statistical Control by Monitoring and Feedback Adjustment
BRANDIMARTE · Numerical Methods in Finance: A MATLAB-Based Introduction
BROWN and HOLLANDER · Statistics: A Biomedical Introduction
BRUNNER, DOMHOF, and LANGER · Nonparametric Analysis of Longitudinal Data in Factorial Experiments
BUCKLEW · Large Deviation Techniques in Decision, Simulation, and Estimation
CAIROLI and DALANG · Sequential Stochastic Optimization
CHAN · Time Series: Applications to Finance
CHATTERJEE and HADI · Sensitivity Analysis in Linear Regression
CHATTERJEE and PRICE · Regression Analysis by Example, *Third Edition*
CHERNICK · Bootstrap Methods: A Practitioner's Guide
CHILÈS and DELFINER · Geostatistics: Modeling Spatial Uncertainty
CHOW and LIU · Design and Analysis of Clinical Trials: Concepts and Methodologies
CLARKE and DISNEY · Probability and Random Processes: A First Course with Applications, *Second Edition*
*COCHRAN and COX · Experimental Designs, *Second Edition*
CONGDON · Bayesian Statistical Modelling
CONOVER · Practical Nonparametric Statistics, *Second Edition*
COOK · Regression Graphics
COOK and WEISBERG · Applied Regression Including Computing and Graphics
COOK and WEISBERG · An Introduction to Regression Graphics
CORNELL · Experiments with Mixtures, Designs, Models, and the Analysis of Mixture Data, *Third Edition*
COVER and THOMAS · Elements of Information Theory
COX · A Handbook of Introductory Statistical Methods
*COX · Planning of Experiments
CRESSIE · Statistics for Spatial Data, *Revised Edition*
CSÖRGŐ and HORVÁTH · Limit Theorems in Change Point Analysis
DANIEL · Applications of Statistics to Industrial Experimentation
DANIEL · Biostatistics: A Foundation for Analysis in the Health Sciences, *Sixth Edition*
*DANIEL · Fitting Equations to Data: Computer Analysis of Multifactor Data, *Second Edition*

*Now available in a lower priced paperback edition in the Wiley Classics Library.

DAVID · Order Statistics, *Second Edition*
*DEGROOT, FIENBERG, and KADANE · Statistics and the Law
DEL CASTILLO · Statistical Process Adjustment for Quality Control
DETTE and STUDDEN · The Theory of Canonical Moments with Applications in Statistics, Probability, and Analysis
DEY and MUKERJEE · Fractional Factorial Plans
DILLON and GOLDSTEIN · Multivariate Analysis: Methods and Applications
DODGE · Alternative Methods of Regression
*DODGE and ROMIG · Sampling Inspection Tables, *Second Edition*
*DOOB · Stochastic Processes
DOWDY and WEARDEN · Statistics for Research, *Second Edition*
DRAPER and SMITH · Applied Regression Analysis, *Third Edition*
DRYDEN and MARDIA · Statistical Shape Analysis
DUDEWICZ and MISHRA · Modern Mathematical Statistics
DUNN and CLARK · Applied Statistics: Analysis of Variance and Regression, *Second Edition*
DUNN and CLARK · Basic Statistics: A Primer for the Biomedical Sciences, *Third Edition*
DUPUIS and ELLIS · A Weak Convergence Approach to the Theory of Large Deviations
*ELANDT-JOHNSON and JOHNSON · Survival Models and Data Analysis
ETHIER and KURTZ · Markov Processes: Characterization and Convergence
EVANS, HASTINGS, and PEACOCK · Statistical Distributions, *Third Edition*
FELLER · An Introduction to Probability Theory and Its Applications, Volume I, *Third Edition,* Revised; Volume II, *Second Edition*
FISHER and VAN BELLE · Biostatistics: A Methodology for the Health Sciences
*FLEISS · The Design and Analysis of Clinical Experiments
FLEISS · Statistical Methods for Rates and Proportions, *Second Edition*
FLEMING and HARRINGTON · Counting Processes and Survival Analysis
FULLER · Introduction to Statistical Time Series, *Second Edition*
FULLER · Measurement Error Models
GALLANT · Nonlinear Statistical Models
GHOSH, MUKHOPADHYAY, and SEN · Sequential Estimation
GIFI · Nonlinear Multivariate Analysis
GLASSERMAN and YAO · Monotone Structure in Discrete-Event Systems
GNANADESIKAN · Methods for Statistical Data Analysis of Multivariate Observations, *Second Edition*
GOLDSTEIN and LEWIS · Assessment: Problems, Development, and Statistical Issues
GREENWOOD and NIKULIN · A Guide to Chi-Squared Testing
GROSS and HARRIS · Fundamentals of Queueing Theory, *Third Edition*
*HAHN · Statistical Models in Engineering
HAHN and MEEKER · Statistical Intervals: A Guide for Practitioners
HALD · A History of Probability and Statistics and their Applications Before 1750
HALD · A History of Mathematical Statistics from 1750 to 1930
HAMPEL · Robust Statistics: The Approach Based on Influence Functions
HANNAN and DEISTLER · The Statistical Theory of Linear Systems
HEIBERGER · Computation for the Analysis of Designed Experiments
HEDAYAT and SINHA · Design and Inference in Finite Population Sampling
HELLER · MACSYMA for Statisticians
HINKELMAN and KEMPTHORNE: · Design and Analysis of Experiments, Volume 1: Introduction to Experimental Design
HOAGLIN, MOSTELLER, and TUKEY · Exploratory Approach to Analysis of Variance
HOAGLIN, MOSTELLER, and TUKEY · Exploring Data Tables, Trends and Shapes

*Now available in a lower priced paperback edition in the Wiley Classics Library.

*HOAGLIN, MOSTELLER, and TUKEY · Understanding Robust and Exploratory Data Analysis
HOCHBERG and TAMHANE · Multiple Comparison Procedures
HOCKING · Methods and Applications of Linear Models: Regression and the Analysis of Variables
HOEL · Introduction to Mathematical Statistics, *Fifth Edition*
HOGG and KLUGMAN · Loss Distributions
HOLLANDER and WOLFE · Nonparametric Statistical Methods, *Second Edition*
HOSMER and LEMESHOW · Applied Logistic Regression, *Second Edition*
HOSMER and LEMESHOW · Applied Survival Analysis: Regression Modeling of Time to Event Data
HØYLAND and RAUSAND · System Reliability Theory: Models and Statistical Methods
HUBER · Robust Statistics
HUBERTY · Applied Discriminant Analysis
HUNT and KENNEDY · Financial Derivatives in Theory and Practice
HUSKOVA, BERAN, and DUPAC · Collected Works of Jaroslav Hajek—with Commentary
IMAN and CONOVER · A Modern Approach to Statistics
JACKSON · A User's Guide to Principle Components
JOHN · Statistical Methods in Engineering and Quality Assurance
JOHNSON · Multivariate Statistical Simulation
JOHNSON and BALAKRISHNAN · Advances in the Theory and Practice of Statistics: A Volume in Honor of Samuel Kotz
JUDGE, GRIFFITHS, HILL, LÜTKEPOHL, and LEE · The Theory and Practice of Econometrics, *Second Edition*
JOHNSON and KOTZ · Distributions in Statistics
JOHNSON and KOTZ (editors) · Leading Personalities in Statistical Sciences: From the Seventeenth Century to the Present
JOHNSON, KOTZ, and BALAKRISHNAN · Continuous Univariate Distributions, Volume 1, *Second Edition*
JOHNSON, KOTZ, and BALAKRISHNAN · Continuous Univariate Distributions, Volume 2, *Second Edition*
JOHNSON, KOTZ, and BALAKRISHNAN · Discrete Multivariate Distributions
JOHNSON, KOTZ, and KEMP · Univariate Discrete Distributions, *Second Edition*
JUREČKOVÁ and SEN · Robust Statistical Procedures: Asymptotics and Interrelations
JUREK and MASON · Operator-Limit Distributions in Probability Theory
KADANE · Bayesian Methods and Ethics in a Clinical Trial Design
KADANE AND SCHUM · A Probabilistic Analysis of the Sacco and Vanzetti Evidence
KALBFLEISCH and PRENTICE · The Statistical Analysis of Failure Time Data
KASS and VOS · Geometrical Foundations of Asymptotic Inference
KAUFMAN and ROUSSEEUW · Finding Groups in Data: An Introduction to Cluster Analysis
KENDALL, BARDEN, CARNE, and LE · Shape and Shape Theory
KHURI · Advanced Calculus with Applications in Statistics
KHURI, MATHEW, and SINHA · Statistical Tests for Mixed Linear Models
KLUGMAN, PANJER, and WILLMOT · Loss Models: From Data to Decisions
KLUGMAN, PANJER, and WILLMOT · Solutions Manual to Accompany Loss Models: From Data to Decisions
KOTZ, BALAKRISHNAN, and JOHNSON · Continuous Multivariate Distributions, Volume 1, *Second Edition*
KOTZ and JOHNSON (editors) · Encyclopedia of Statistical Sciences: Volumes 1 to 9 with Index
KOTZ and JOHNSON (editors) · Encyclopedia of Statistical Sciences: Supplement Volume

*Now available in a lower priced paperback edition in the Wiley Classics Library.

KOTZ, READ, and BANKS (editors) · Encyclopedia of Statistical Sciences: Update Volume 1

KOTZ, READ, and BANKS (editors) · Encyclopedia of Statistical Sciences: Update Volume 2

KOVALENKO, KUZNETZOV, and PEGG · Mathematical Theory of Reliability of Time-Dependent Systems with Practical Applications

LACHIN · Biostatistical Methods: The Assessment of Relative Risks

LAD · Operational Subjective Statistical Methods: A Mathematical, Philosophical, and Historical Introduction

LAMPERTI · Probability: A Survey of the Mathematical Theory, *Second Edition*

LANGE, RYAN, BILLARD, BRILLINGER, CONQUEST, and GREENHOUSE · Case Studies in Biometry

LARSON · Introduction to Probability Theory and Statistical Inference, *Third Edition*

LAWLESS · Statistical Models and Methods for Lifetime Data

LAWSON · Statistical Methods in Spatial Epidemiology

LE · Applied Categorical Data Analysis

LE · Applied Survival Analysis

LEE · Statistical Methods for Survival Data Analysis, *Second Edition*

LePAGE and BILLARD · Exploring the Limits of Bootstrap

LEYLAND and GOLDSTEIN (editors) · Multilevel Modelling of Health Statistics

LIAO · Statistical Group Comparison

LINDVALL · Lectures on the Coupling Method

LINHART and ZUCCHINI · Model Selection

LITTLE and RUBIN · Statistical Analysis with Missing Data

LLOYD · The Statistical Analysis of Categorical Data

MAGNUS and NEUDECKER · Matrix Differential Calculus with Applications in Statistics and Econometrics, *Revised Edition*

MALLER and ZHOU · Survival Analysis with Long Term Survivors

MALLOWS · Design, Data, and Analysis by Some Friends of Cuthbert Daniel

MANN, SCHAFER, and SINGPURWALLA · Methods for Statistical Analysis of Reliability and Life Data

MANTON, WOODBURY, and TOLLEY · Statistical Applications Using Fuzzy Sets

MARDIA and JUPP · Directional Statistics

MASON, GUNST, and HESS · Statistical Design and Analysis of Experiments with Applications to Engineering and Science

McCULLOCH and SEARLE · Generalized, Linear, and Mixed Models

McFADDEN · Management of Data in Clinical Trials

McLACHLAN · Discriminant Analysis and Statistical Pattern Recognition

McLACHLAN and KRISHNAN · The EM Algorithm and Extensions

McLACHLAN and PEEL · Finite Mixture Models

McNEIL · Epidemiological Research Methods

MEEKER and ESCOBAR · Statistical Methods for Reliability Data

MEERSCHAERT and SCHEFFLER · Limit Distributions for Sums of Independent Random Vectors: Heavy Tails in Theory and Practice

*MILLER · Survival Analysis, *Second Edition*

MONTGOMERY, PECK, and VINING · Introduction to Linear Regression Analysis, *Third Edition*

MORGENTHALER and TUKEY · Configural Polysampling: A Route to Practical Robustness

MUIRHEAD · Aspects of Multivariate Statistical Theory

MURRAY · X-STAT 2.0 Statistical Experimentation, Design Data Analysis, and Nonlinear Optimization

MYERS and MONTGOMERY · Response Surface Methodology: Process and Product Optimization Using Designed Experiments, *Second Edition*

---

*Now available in a lower priced paperback edition in the Wiley Classics Library.

MYERS, MONTGOMERY, and VINING · Generalized Linear Models. With Applications in Engineering and the Sciences
NELSON · Accelerated Testing, Statistical Models, Test Plans, and Data Analyses
NELSON · Applied Life Data Analysis
NEWMAN · Biostatistical Methods in Epidemiology
OCHI · Applied Probability and Stochastic Processes in Engineering and Physical Sciences
OKABE, BOOTS, SUGIHARA, and CHIU · Spatial Tesselations: Concepts and Applications of Voronoi Diagrams, *Second Edition*
OLIVER and SMITH · Influence Diagrams, Belief Nets and Decision Analysis
PANKRATZ · Forecasting with Dynamic Regression Models
PANKRATZ · Forecasting with Univariate Box-Jenkins Models: Concepts and Cases
*PARZEN · Modern Probability Theory and Its Applications
PEÑA, TIAO, and TSAY · A Course in Time Series Analysis
PIANTADOSI · Clinical Trials: A Methodologic Perspective
PORT · Theoretical Probability for Applications
POURAHMADI · Foundations of Time Series Analysis and Prediction Theory
PRESS · Bayesian Statistics: Principles, Models, and Applications
PRESS and TANUR · The Subjectivity of Scientists and the Bayesian Approach
PUKELSHEIM · Optimal Experimental Design
PURI, VILAPLANA, and WERTZ · New Perspectives in Theoretical and Applied Statistics
PUTERMAN · Markov Decision Processes: Discrete Stochastic Dynamic Programming
*RAO · Linear Statistical Inference and Its Applications, *Second Edition*
RENCHER · Linear Models in Statistics
RENCHER · Methods of Multivariate Analysis, *Second Edition*
RENCHER · Multivariate Statistical Inference with Applications
RIPLEY · Spatial Statistics
RIPLEY · Stochastic Simulation
ROBINSON · Practical Strategies for Experimenting
ROHATGI and SALEH · An Introduction to Probability and Statistics, *Second Edition*
ROLSKI, SCHMIDLI, SCHMIDT, and TEUGELS · Stochastic Processes for Insurance and Finance
ROSS · Introduction to Probability and Statistics for Engineers and Scientists
ROUSSEEUW and LEROY · Robust Regression and Outlier Detection
RUBIN · Multiple Imputation for Nonresponse in Surveys
RUBINSTEIN · Simulation and the Monte Carlo Method
RUBINSTEIN and MELAMED · Modern Simulation and Modeling
RYAN · Modern Regression Methods
RYAN · Statistical Methods for Quality Improvement, *Second Edition*
SALTELLI, CHAN, and SCOTT (editors) · Sensitivity Analysis
*SCHEFFE · The Analysis of Variance
SCHIMEK · Smoothing and Regression: Approaches, Computation, and Application
SCHOTT · Matrix Analysis for Statistics
SCHUSS · Theory and Applications of Stochastic Differential Equations
SCOTT · Multivariate Density Estimation: Theory, Practice, and Visualization
*SEARLE · Linear Models
SEARLE · Linear Models for Unbalanced Data
SEARLE · Matrix Algebra Useful for Statistics
SEARLE, CASELLA, and McCULLOCH · Variance Components
SEARLE and WILLETT · Matrix Algebra for Applied Economics
SEBER · Linear Regression Analysis
SEBER · Multivariate Observations
SEBER and WILD · Nonlinear Regression
SENNOTT · Stochastic Dynamic Programming and the Control of Queueing Systems

*Now available in a lower priced paperback edition in the Wiley Classics Library.

*SERFLING · Approximation Theorems of Mathematical Statistics
SHAFER and VOVK · Probability and Finance: It's Only a Game!
SMALL and McLEISH · Hilbert Space Methods in Probability and Statistical Inference
STAPLETON · Linear Statistical Models
STAUDTE and SHEATHER · Robust Estimation and Testing
STOYAN, KENDALL, and MECKE · Stochastic Geometry and Its Applications, *Second Edition*
STOYAN and STOYAN · Fractals, Random Shapes and Point Fields: Methods of Geometrical Statistics
STYAN · The Collected Papers of T. W. Anderson: 1943–1985
SUTTON, ABRAMS, JONES, SHELDON, and SONG · Methods for Meta-Analysis in Medical Research
TANAKA · Time Series Analysis: Nonstationary and Noninvertible Distribution Theory
THOMPSON · Empirical Model Building
THOMPSON · Sampling, *Second Edition*
THOMPSON · Simulation: A Modeler's Approach
THOMPSON and SEBER · Adaptive Sampling
TIAO, BISGAARD, HILL, PEÑA, and STIGLER (editors) · Box on Quality and Discovery: with Design, Control, and Robustness
TIERNEY · LISP-STAT: An Object-Oriented Environment for Statistical Computing and Dynamic Graphics
TSAY · Analysis of Financial Time Series
UPTON and FINGLETON · Spatial Data Analysis by Example, Volume II: Categorical and Directional Data
VAN BELLE · Statistical Rules of Thumb
VIDAKOVIC · Statistical Modeling by Wavelets
WEISBERG · Applied Linear Regression, *Second Edition*
WELSH · Aspects of Statistical Inference
WESTFALL and YOUNG · Resampling-Based Multiple Testing: Examples and Methods for *p*-Value Adjustment
WHITTAKER · Graphical Models in Applied Multivariate Statistics
WINKER · Optimization Heuristics in Economics: Applications of Threshold Accepting
WONNACOTT and WONNACOTT · Econometrics, *Second Edition*
WOODING · Planning Pharmaceutical Clinical Trials: Basic Statistical Principles
WOOLSON and CLARKE · Statistical Methods for the Analysis of Biomedical Data, *Second Edition*
WU and HAMADA · Experiments: Planning, Analysis, and Parameter Design Optimization
YANG · The Construction Theory of Denumerable Markov Processes
*ZELLNER · An Introduction to Bayesian Inference in Econometrics
ZHOU, OBUCHOWSKI, and McCLISH · Statistical Methods in Diagnostic Medicine

*Now available in a lower priced paperback edition in the Wiley Classics Library.